Archaeology and Tourism

TOURISM AND CULTURAL CHANGE

Series Editors: Professor Mike Robinson, *Ironbridge International Institute for Cultural Heritage, University of Birmingham, UK* and Professor Alison Phipps, *University of Glasgow, Scotland, UK*

Understanding tourism's relationships with culture(s) and vice versa, is of ever-increasing significance in a globalising world. TCC is a series of books that critically examine the complex and ever-changing relationship between tourism and culture(s). The series focuses on the ways that places, peoples, pasts, and ways of life are increasingly shaped/transformed/created/packaged for touristic purposes. The series examines the ways tourism utilises/makes and re-makes cultural capital in its various guises (visual and performing arts, crafts, festivals, built heritage, cuisine etc.) and the multifarious political, economic, social and ethical issues that are raised as a consequence. Theoretical explorations, research-informed analyses, and detailed historical reviews from a variety of disciplinary perspectives are invited to consider such relationships.

All books in this series are externally peer-reviewed.

Full details of all the books in this series and of all our other publications can be found on http://www.channelviewpublications.com, or by writing to Channel View Publications, St Nicholas House, 31–34 High Street, Bristol BS1 2AW, UK.

TOURISM AND CULTURAL CHANGE: 55

Archaeology and Tourism

Touring the Past

Edited by
**Dallen J. Timothy and
Lina G. Tahan**

CHANNEL VIEW PUBLICATIONS
Bristol • Blue Ridge Summit

DOI https://doi.org/10.21832/TIMOTH7567
Library of Congress Cataloging in Publication Data
A catalog record for this book is available from the Library of Congress.
Names: Timothy, Dallen J., editor, author. | Tahan, Lina G. - editor, author.
Title: Archaeology and Tourism: Touring the Past/Edited by Dallen J. Timothy and Lina G. Tahan.
Description: Blue Ridge Summit; Bristol: Channel View Publications, 2020. | Series: Tourism and Cultural Change: 55 | Includes bibliographical references and index. | Summary: "This book provides a global examination of the relationships between archaeology and tourism. It offers a critical analysis of current issues and implications from both tourism and archaeological perspectives. It will be useful for students, researchers and practitioners in tourism, archaeology, cultural heritage management and anthropology"— Provided by publisher.
Identifiers: LCCN 2020002504 (print) | LCCN 2020002505 (ebook) | ISBN 9781845417550 (paperback) | ISBN 9781845417567 (hardback) | ISBN 9781845417574 (pdf) | ISBN 9781845417581 (epub) | ISBN 9781845417598 (kindle edition) Subjects: LCSH: Heritage tourism. | Historic sites. | Antiquities. | Archaeology. | Historic preservation.
Classification: LCC G156.5.H47 A74 2020 (print) | LCC G156.5.H47 (ebook) | DDC 910.68—dc23 LC record available at https://lccn.loc.gov/2020002504
LC ebook record available at https://lccn.loc.gov/2020002505

British Library Cataloguing in Publication Data
A catalogue entry for this book is available from the British Library.

ISBN-13: 978-1-84541-756-7 (hbk)
ISBN-13: 978-1-84541-755-0 (pbk)

Channel View Publications
UK: St Nicholas House, 31–34 High Street, Bristol BS1 2AW, UK.
USA: NBN, Blue Ridge Summit, PA, USA.

Website: www.channelviewpublications.com
Twitter: Channel_View
Facebook: https://www.facebook.com/channelviewpublications
Blog: www.channelviewpublications.wordpress.com

Copyright © 2020 Dallen J. Timothy, Lina G. Tahan and the authors of individual chapters.

All rights reserved. No part of this work may be reproduced in any form or by any means without permission in writing from the publisher.

The policy of Multilingual Matters/Channel View Publications is to use papers that are natural, renewable and recyclable products, made from wood grown in sustainable forests. In the manufacturing process of our books, and to further support our policy, preference is given to printers that have FSC and PEFC Chain of Custody certification. The FSC and/or PEFC logos will appear on those books where full certification has been granted to the printer concerned.

Typeset by Nova Techset Private Limited, Bengaluru and Chennai, India.

Contents

Tables and Figures		vii
Contributors		ix
1	Archaeology and Tourism: Consuming, Managing and Protecting the Human Past *Dallen J. Timothy and Lina G. Tahan*	1
2	Archaeologists and Tourism: Symbiosis or Contestation? *Laurence Gillot*	26
3	Tourism and the Economic Value of Archaeology *Paul Burtenshaw*	41
4	Privatization, Archaeology and Tourism *Işılay Gürsu*	54
5	Marketing Archaeological Heritage for Tourism *Alan Fyall, Anna Leask and Sarah B. Barber*	69
6	Archaeological Heritage and Volunteer Tourism *Dallen J. Timothy*	87
7	Archaeology and Religious Tourism: Sacred Sites, Rituals, Sharing the *Baraka* and Tourism Development *Nour Farra-Haddad*	106
8	Archaeological Destruction and Tourism: Sites, Sights, Rituals and Narratives *Lina G. Tahan*	121
9	Plundering the Past: Tourism and the Illicit Trade in Archaeological Remains *Dallen J. Timothy*	134
10	Protecting the Archaeological Past in the Face of Tourism Demand *Jennifer P. Mathews*	152

11 Interpreting the Past: Telling the Archaeological Story
    to Visitors   167
    Sue Hodges

12 Archaeology, Nationalism and Politics: The Need for Tourism   186
    Gai Jorayev

13 Understanding Perspectives on Archaeology and Tourism   205
    Dallen J. Timothy and Lina G. Tahan

    Index   224

Tables and Figures

Tables

Table 6.1	Examples of archaeological excavations soliciting volunteer tourists for 2019	99
Table 6.2	A sample of companies that sell or promote archaeology-oriented volunteer vacations	100
Table 9.1	Supply-side relationships between tourism and antiquities consumption	141

Figures

Figure 1.1	The Great Wall of China is an iconic symbol of tourism in China	5
Figure 1.2	Tourists visiting the active excavation site of Çatalhöyük, Turkey	6
Figure 1.3	Handball courts at the ancient Maya city of Chichen Itza provide archaeological evidence of the ancient development of sport	11
Figure 1.4	The graffiti left behind by Medieval pilgrims in the Holy Land has become part of the heritage appeal of some attractions	13
Figure 3.1	Section of the Bronze Age linear cemetery with Kilmartin Village in the background	45
Figure 3.2	Stone circle at Templewood, Kilmartin Glen	46
Figure 5.1	The Mesoamerican ballcourt at Copalita. These structures are popular architectural features among tourists and often a highlight of site tours in Mexico	77
Figure 5.2	Interior of the site museum at the Parque Eco-Arqueologico Copalita containing reproductions of archaeological finds from elsewhere in Oaxaca state	79

Figure 5.3	The 'History Bug' marketing campaign	82
Figure 6.1	University and high school student 'tourists' volunteering on a Fremont Indian archaeological dig in southern Utah, USA	93
Figure 6.2	Archaeology enthusiasts volunteer their vacation time to help clean and catalogue artifacts at a museum in Philadelphia, USA	95
Figure 7.1	The Afqa Cave in rural Jbeil-Byblos, Lebanon, with the remains of the Roman Temple of Venus, shaken by earthquakes and eroded by time	109
Figure 7.2	In the village of Cana, Lebanon, is the 'Site of the Statuary', where rock carvings date back many centuries	116
Figure 8.1	Ayodhya seen from the Ghaghara River, Uttar Pradesh, with the mosque depicted in the upper left. Coloured etching by William Hodges, 1785	124
Figure 8.2	SOLIDERE wanted to create a 'garden of forgiveness' using the Roman basilica	127
Figure 8.3	Contestation between developers and archaeologists has resulted in a stalemate in one of Beirut's archaeological areas	128
Figure 9.1	This antiquities shop in Jerusalem provides certified and legal sales of archaeological artefacts to tourists	143
Figure 13.1	Many countries, including China, are especially eager to extend the UNESCO 'brand' to as many archaeological sites as possible	206
Figure 13.2	Special-interest displays, such as this one of the Bredgar Hoard of Roman coins in the British Museum, appeal to niche markets	210
Figure 13.3	Spas were ubiquitous during the Roman Empire, and many, such as Terme di Caracalla in Rome, now serve as important archaeological attractions	212
Figure 13.4	Like this setting at Longmen Caves, China, QR Codes and other technologies have become a commonplace interpretive tool in recent years	218
Figure 13.5	Augmented reality helps 'reconstruct' ruins digitally or allows visitors to see how archaeological sites might have looked during different periods of history	218

Contributors

Sarah Barber is Associate Professor of Anthropology and with the National Center for Integrated Coastal Research at the University of Central Florida. She is a Mesoamerican archaeologist specializing in the origins and organization of early urban societies, including the roles played by religion, long-distance exchange and coastal ecosystems in those processes. She has directed multiple field research projects in coastal Oaxaca, Mexico, and Florida, United States. She is the co-editor of *Religion and Politics in the Ancient Americas*, published by Routledge (2018), and has published her results in journals including the *Journal of Archaeological Science, Journal of Archaeological Science: Reports, Current Anthropology, Ancient Mesoamerica* and the *Archaeological Papers of the American Anthropological Association*. Her research has been funded by the National Science Foundation, the Historic Society, the National Geographic Society and Argonne National Laboratories.

Paul Burtenshaw is an archaeologist and independent researcher and consultant on cultural heritage and sustainable development. Between 2014 and 2019, he was the Director of Projects at the Sustainable Preservation Initiative. The Sustainable Preservation Initiative is a US-based non-profit that develops sustainable community economic enterprises associated with cultural heritage. Burtenshaw's PhD research examined the concepts around the 'economic values' of archaeology, how they are used to mobilize value for, and justify the preservation of, archaeology at international, national and local levels, and the methods archaeologists can use to measure economic value. He has been a Visiting Research Fellow at the Centre for British Research in the Levant, Amman, as well as co-organizer (with Peter Gould) of the Archaeology and Economic Development Conference held at UCL in September 2012.

Nour Farra-Haddad is a religious anthropologist. She is a senior researcher and teaches at St-Joseph University of Beirut, at the Lebanese University and at the American University of Science and Technology (Lebanon). She also manages her own travel consultancy company, NEOS. She has a PhD in religious anthropology, a master's degree in anthropology and two bachelor's degrees in archaeology and sociology. She holds a diploma in tour guiding from the Ministry of Tourism and she is a founding member of the Association for the Development of Pilgrimages and Religious Tourism (APL). She is the author of two tourist guides – 'Wiz Kids' and 'Eco Lebanon: Nature and Rural Tourism' – and she recently developed the first application promoting interreligious tourism in Lebanon: 'Holy Lebanon'.

Alan Fyall is Associate Dean of Academic Affairs and Visit Orlando Endowed Chair of Tourism Marketing at the Rosen College of Hospitality Management, and is a member of the University of Central Florida's National Center for Integrated Coastal Research. His current research interests relate to coastal tourism and destination resilience in Florida and the Caribbean. Alan currently teaches International Tourism Management and Destination Marketing & Management and to date has examined 27 PhDs in the UK, India, France, South Africa, Australia, Hong Kong and Malaysia. He has organized a number of international conferences and workshops for academic, professional and governmental audiences and is frequently invited to deliver keynote addresses. He is the editor of Elsevier's *Journal of Destination Marketing & Management* and sits on the editorial boards of many leading journals.

Laurence Gillot is a Senior Lecturer at the University of Paris 7 (Paris Diderot) and is co-responsible for the master's degree 'City, Architecture, Heritage'. She teaches the history and sociology of heritage. Her research is at the crossroads of various scientific disciplines (history, archaeology, geography and sociology) and concerns the relationship of present societies to the past and its material traces. She is studying in particular the representations and uses of antiquity in the Arab-Muslim region (Morocco, Syria). Consequently, she is interested in issues relating to the development of archaeological tourism in these countries.

Işılay Gürsu received her PhD in Cultural Heritage Management from IMT Institute for Advanced Studies, Lucca, Italy, in 2013. She joined the British Institute at Ankara in January 2013, where she works as a Cultural Heritage Management Research Fellow. She has been involved in several heritage management and archaeology research projects in Turkey and currently is involved in the Safeguarding Archaeological Assets of Turkey project. She completed a tourism administration degree from Boğaziçi University and a master's degree at Koç University in Anatolian Civilizations and Cultural Heritage Management before completing her PhD in Italy. Her scholarly interests include understanding the contemporary relationships between society and archaeology.

Sue Hodges is an historian from Melbourne, Australia, with extensive experience in the fields of history, heritage interpretation, sustainable tourism, capacity building, placemaking and museum and exhibition development. She has a BA Hons (History, English) and MA (Public History) and is currently President of the ICOMOS International Scientific Committee on the Interpretation and Presentation of Cultural Heritage Sites (ICIP), a Member of the ICOMOS Advisory Committee and an International Expert Member of the Foundazione Romualdo Del Bianco. Sue was an invited expert speaker at the 40th and 41st Sessions of the World Heritage Committee, President of Interpretation Australia from 2010 to 2013 and an Executive Committee Member of Australia ICOMOS from 2012 to 2015. Her business, SHP, operates in Australia and internationally.

Gai Jorayev, PhD, holds degrees in tourism and archaeological heritage management and currently works at UCL Institute of Archaeology. He is a Degree Programme Co-ordinator for the MA in Managing Archaeological Sites and works as part of the research teams focusing on two main areas of modern-day heritage management and digital heritage. He recently carried out projects on post-Soviet space and Africa and the European Union, and he collaborates closely with organizations such as UNESCO, ICOMOS, UNWTO and international development agencies. He is experienced in spatial analysis and modelling, and uses it for public outreach, interpretation and education. Creating open-access heritage inventory systems to monitor change and to inform decisions in heritage management and cultural tourism among others is among the key areas of his current research.

Jennifer Mathews was trained as an archaeologist and is a Professor of Anthropology at Trinity University. She studies ancient and historical Maya archaeology, as well as issues of sustainability and tourism. She received her PhD in anthropology from the University of California at Riverside and has been conducting fieldwork in Mexico since 1993. Her books include three edited volumes: *Quintana Roo Archaeology*, *Lifeways in the Northern Maya Lowlands: New Approaches to Archaeology in the Yucatán Peninsula* and *The Value of Things: Prehistoric to Contemporary Commodities in the Maya Region*. She also published *Chicle: Chewing Gum of the Americas: From the Ancient Maya to William Wrigley* and *Sugarcane and Rum: The Bittersweet History of Labor and Life on the Yucatán Peninsula*. She was named the 2019 recipient of the Dr. and Mrs. Z.T. Scott Faculty Fellowship.

Anna Leask (PhD) is Professor of Tourism Management at Edinburgh Napier University, UK. Her teaching and research interests lie principally in the areas of visitor attraction management, heritage tourism and destination management. She has held several external international positions, including Visiting Professor at Wakayama University in Japan

(2016–2019) and as external lecturer at the University of Corsica (2019). Anna's recent research has focused on how visitor attractions can engage with Generation Y visitors and employees, with primary research being conducted in the UK and Asia. Her most recent research relates to identifying the enablers and barriers for older visitors to engage with museums and other visitor attractions, with several ongoing projects on dementia-friendly destinations.

Lina G. Tahan completed a PhD in archaeology and museum studies from the University of Cambridge, United Kingdom. She is currently an affiliated scholar at the department of Archaeology, University of Cambridge and until recently a university lecturer at the Department of Archaeology at the Lebanese University. Her research and teaching interests relate to: (1) representation issues within Middle Eastern museums' collections, exhibitions and visitors; (2) the history of collections and museum development in Lebanon within the political context; (3) the definition of heritage in Lebanon and how this links to identity formation and the creation of a sense of place; (4) the role of museums in fostering understanding in divided societies. She is an active member of the International Council of Museums (ICOM) and a board member of the International Committee for Ethical Dilemmas. She is mainly working on promoting museums in the Arab World.

Dallen J. Timothy is a human geographer by training and Professor of Community Resources and Development at Arizona State University and Senior Sustainability Scientist at the Wrigley Global Institute of Sustainability. He also holds several visiting professorships in Spain and China and is a Senior Research Associate at the University of Johannesburg, South Africa. Professor Timothy is Editor-in-Chief of the *Journal of Heritage Tourism*, and serves on the editorial boards of 24 academic journals. He has many research interests, including heritage management, religious tourism, food as cultural heritage, indigenous people, international borders and their effects on tourism, migration and globalization processes. He has worked in various capacities in more than 100 countries, and has ongoing research projects on several tourism-related topics in North America, Asia, Africa, Europe and the Middle East.

1 Archaeology and Tourism: Consuming, Managing and Protecting the Human Past

Dallen J. Timothy and Lina G. Tahan

Introduction

Human beings have a long history of mobility for many social and economic purposes, including hunting and fishing, trade, warfare, celebrations and religious pilgrimages. Some archaeologists believe the Turkish site of Göbekli Tepe (10th–8th millennium BCE) to be one of the earliest spiritual gathering places or centers of worship ever discovered. Evidence of religious pilgrimages has also been found from the Vedic age in northern India (c. 1500–500 BCE) during the early stages of Hinduism. Pilgrimages thrived during the Middle Ages in Asia and Europe, although portrayals of this as an early form of tourism have focused largely on the movement of Christians between Europe and Jerusalem, and throughout the lands of the Bible, until the 16th-century Christian Reformation prohibited pilgrimage travel for many Europeans.

Other types of tourism also have a long history. There are records of pleasure travel and 'sightseeing' in Egypt as early as 1500 years BCE (Casson, 1994). During antiquity and the Middle Ages, social elites traveled for 'holiday-making'. Many ancient accounts suggest that nobility, merchants, traders and the aristocracy during antiquity traveled to see sites and places that were already considered quite old. The seven wonders of the ancient world were important destinations during the Greek and Roman empires. In fact, the earliest Greek guidebooks included descriptions and travelers' reviews of the Egyptian Pyramids, the Temple of Artemis at Ephesus, the Colossus of Rhodes, the Mausoleum of Mausolus, the Ishtar Gate and the Statue of Zeus at Olympia (Timothy, 2011).

As well, the Grand Tour of Europe (17th–19th centuries) was an activity wherein young aristocratic males traveled to various European destinations on set itineraries to view great works of art, historic cities, ancient ruins, and to learn from the artistic masters. This became a right of passage for many youth of the upper classes and is frequently cited as the early forerunner to modern-day tourism (Towner, 1985). The destination foci of the Grand Tour, the sightseeing and holidaymaking activities during classical antiquity and the Middle Ages, and the contemporary manifestations of modern tourism, as exemplified by Thomas Cook in 19th-century Great Britain, almost always pointed to archaeological sites and other parts of the historic environment.

Even today, archaeology remains one of the most ubiquitous assets of present-day tourism, and many worldwide destinations depend largely, or almost entirely, on archaeological remains and other heritage for their tourism economies. While cultural heritage covers a very broad range of resources, of particular interest in this book is built and tangible heritage, namely archaeology. Although heritage and archaeology are not synonymous, they are overlapping concepts; in fact, the archaeological record is part of the broader realm of heritage (Emerick, 2009). This introductory chapter provides an overview of many of the issues in the crossover between tourism and archaeology and sets the conceptual tone for the remainder of the book. It first examines the relationship between archaeology and heritage, suggesting that they are not synonymous but overlapping. The chapter then examines several of the many relationships between archaeology and tourism, and highlights the contents of the book.

Archaeology and Heritage

Archaeology is the scientific field that studies humankind's past activities by analyzing remnants of material culture. It utilizes techniques, concepts, theories and interpretive tools from the social sciences, physical sciences and humanities. Archaeologists seek to understand past and present human behavior, the origins of humans and their cultures, and the ways in which societies develop over time (Ashmore & Sharer, 2014). Archaeologists use manufactured tools, bones, burial sites, food remains, buildings and other artifacts to discover how people lived in the past and to draw parallels to how we live today. Their work is typically done in three main phases: site surveys to learn as much as possible about the area under study, excavations to uncover buried cultural artifacts or assessments of uncovered buildings and artifacts, and data analysis and publishing the findings.

Site surveys may involve remote sensing to analyze satellite imagery, aerial photographs and drone images, as well as surface surveys. This often entails soil sampling, 'shovel tests', radar and laser checks, metal detecting and other similar exercises. Excavations involve digging layers of strata, artifact discovery, measuring and recording contexts, photographing, sifting soil and cleaning. Data analysis requires researchers to catalogue and compare the results with previous findings; artifacts are also dated and their compositions studied. Many different tests are available to evaluate the biotic and abiotic composition of artifacts and estimate their ages (Ashmore & Sharer, 2014).

Contrary to popular belief, not all archaeologists or archaeological studies utilize buried artifacts in their quests for knowledge. Many also analyze historic buildings above ground and their environs, landscapes and settings to understand past social and cultural contexts. All material remnants of human civilizations are important parts of the archaeological record. In fact, although mainstream archaeology continues to use the material past as scientific evidence and discovery, some archaeologists are increasingly interested in intangible culture as a means of understanding the broader cultural context of archaeological remains (Akagawa & Smith, 2019; Carman, 2009; Smith & Akagawa, 2009).

There is a wide range of sub-disciplines in archaeology. These are frequently classified by geographical/regional specialization (e.g. Near Eastern archaeology), particular cultures or civilizations (e.g. Assyriology), chronological concentrations (e.g. Neolithic archaeology), specific themes (e.g. Biblical archaeology), methods (e.g. carbon dating), purposes (e.g. rescue archaeology) or materials (e.g. stone tools). Although not all of them are noted here, there are many other ways of categorizing archaeological specialties.

Concerns over protecting the archaeological record led to the establishment of a specific field known as cultural resource management (CRM) or cultural heritage management (CHM) during the 1970s, with archaeology being among the most important tools used by CHM specialists (Emerick, 2009). CRM/CHM derived originally from the subfield of rescue archaeology and is primarily concerned with the protection, documentation and assessment, curation, interpretation, preservation and restoration of archaeological remains. More recently, it includes efforts to protect and interpret intangible culture. This subfield of archaeology also draws heavily on history, anthropology, geography, and ecology to understand how best to analyze and protect the built environment and intangible heritage. The employment of CRM as a professional field also entails working with archaeology consumers, including tourists.

Public archaeology, or community archaeology, is a way of practicing the science that is 'by the people, for the people'. While community archaeology has existed in one form or another for decades (e.g. volunteer archaeology), the term and its practice became particularly popular during the 1970s in the United States, the United Kingdom and other areas of Europe and the Middle East. While it initially meant publicly funded explorations, the term has since come to represent an approach to archaeology that democratizes heritage by engaging the public in archaeological work through participation in excavations and building assessments, tours of sites and digs, public lectures, interpretive programs and archaeological site-oriented events and activities.

Through these outreach actions, the archaeological record becomes better embedded in the community with the aim of stimulating public awareness and interest in heritage, increasing recognition for the need to protect archaeological resources, and helping people connect to their own heritage (Moshenska, 2017). This is especially important for descendent communities, such as indigenous peoples or diasporic groups, who might recognize the value of archaeology in connecting them with their ancestors and deepening their sense of place and rootedness (Davidson & Brandon, 2012). These participatory practices are also viewed as an important way to decolonize archaeology (Tahan, 2010a), which traditionally had been done in a top-down manner by the colonists largely for the good of the colonial metropoles.

This book is first and foremost about the relationships between tourism and archaeology. We recognize that archaeology and heritage are not synonymous, although we do acknowledge that archaeology and its practices and discoveries are a salient part of the much broader domain of heritage and have been considered such for many years (Watson, 2009). Heritage has been variously defined as the present-day use of the past and how modern societies value the past, both its tangible and intangible manifestations (Emerick, 2009; Graham *et al.*, 2000; Timothy, 2011).

Waterton and Smith (2009) have suggested that heritage is more fluid than archaeology, that heritage is a cultural process rather than a measurable 'thing'. Thus, archaeological findings are objectively verified phenomena, whereas heritage reflects more dynamism, subjectivity and negotiable interpretations that may exclude certain communities and elements of the past while including others. This distinction is critical, because for archaeology purists, the vagaries, subjectivity and manipulation of heritage defile the purity of archaeology as the singular and accurate interpretation of material culture (Watson, 2009). From this perspective, then, archaeology itself alone is not heritage, but its use and the social 'collectivism' surrounding it may be manifestations of heritage (Fouseki, 2009), particularly in relation to how archaeology provides the fodder for the development (and manipulation) of popular memory, race and nationhood (Hodder, 2012; Watson, 2009; Wilson, 2009).

As previously noted in relation to CRM/CHM, many contemporary archaeologists study the broader notion of heritage to understand the human past more holistically and within broader sociocultural, economic, political and historical contexts. For the purposes of this book, archaeology and heritage are not synonymous. However, the archaeological record as it is used today is part of a long tradition of conflating heritage and archaeology within the cultural industries and in the field of cultural heritage management (Watson, 2009). Thus, the use of archaeology can be seen as part of the broader heritage movement. Although the focus of this book is archaeology, concepts related to other aspects of heritage manifest as well in a variety of settings that are highlighted throughout the book. The very utilization of archaeology and its findings by the tourism industry by definition reflects the heritagization process and renders them a consumable heritage commodity.

Archaeology and Tourism: Relationships and Perspectives

Tourism has several direct relationships with archaeology, but perhaps the most obvious one is cultural artifacts as regional assets for tourism. Many worldwide destinations boast of their archaeological heritage in their marketing activities and branding efforts, where iconic national symbols are imbued with images of famous ancient monuments (Holtorf, 2007). For example, tourism in India is nearly always associated with the Taj Mahal. Peru's tourism is closely attached to images of Machu Picchu and Jordan's tourism is linked to Petra, just as China's tourism is aligned with the Great Wall (Figure 1.1). While tourism that is composed largely of visits to historic sites and archaeological parks is part of the broader concept of heritage tourism, or cultural tourism, several scholars have begun examining it as a unique niche form of tourism that focuses specifically on archaeological localities, ruins and remnants, so that it is more narrowly defined (Babalola & Ajekigbe, 2007; Giraudo & Porter, 2010; Li

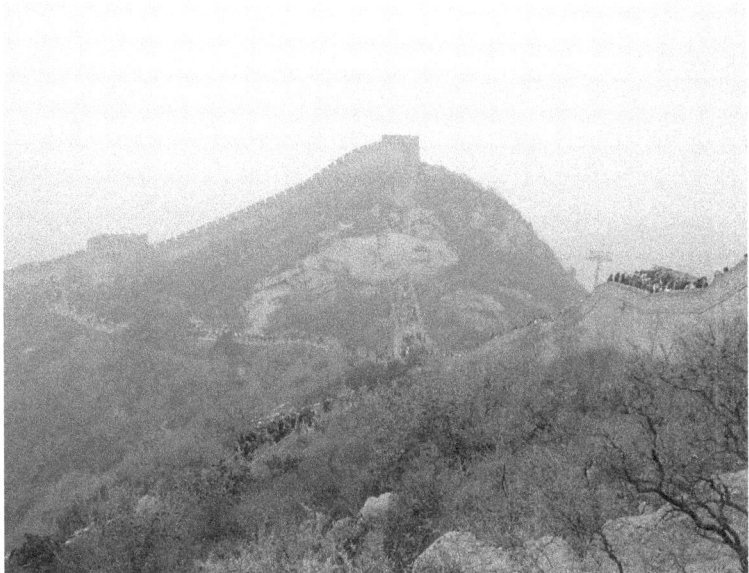

Figure 1.1 The Great Wall of China is an iconic symbol of tourism in China (Photo: Dallen J. Timothy)

& Qian, 2017; Wurz & van der Merwe, 2005) than the broader notion of heritage tourism, which also includes living cultures (Timothy, 2011).

The most common archaeology-related heritage assets that form the tourism product include ruins and archaeological dig sites, ancient monuments, historic buildings, museums, industrial archaeology and interpretive centers. Most tourists see what has already been excavated, restored and preserved and often appears as part of an archaeological park. At other sites, digs are in progress, which enables tourists to see the activities of archaeologists and learn from the scientific process (Ramsey & Everitt, 2008) (Figure 1.2). Archaeology-based tourism occurs in a wide range of physical settings, including national parks, national monuments, archaeological parks and active dig sites. Archaeological work is happening nearly everywhere – wherever there has been past human activity. While every country has an archaeological record, or at least cultural remains, several countries have become famous archaeological locations and subsequently famous destinations for archaeology enthusiasts. Geographic scale, or reach, is an important consideration in this regard. A handful of countries are home to some of the world's best-known archaeological icons. Among these are Egypt, Turkey, Greece, Italy, Spain, China, Palestine, the United Kingdom, Cambodia, Thailand and India. The Roman Forum is one of Italy's tourism claims to fame. The same is true of Stonehenge in England and Angkor Wat in Cambodia. However, even the smallest countries have archaeological remains that are an important part of their national

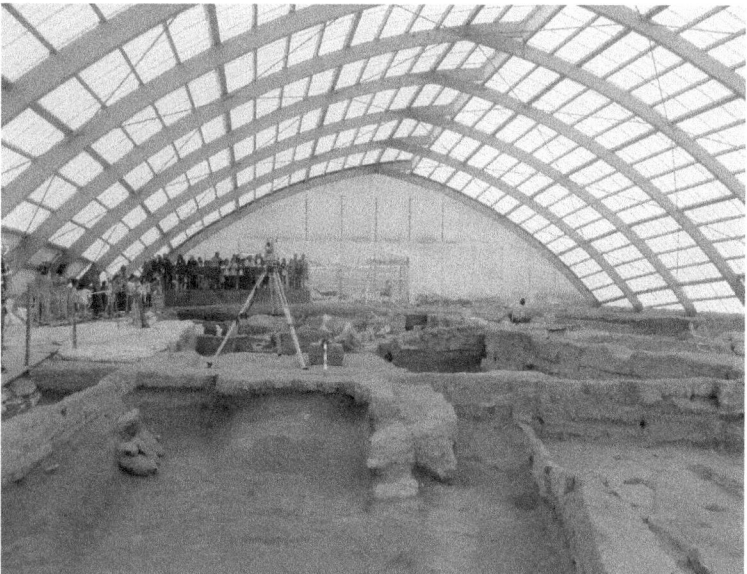

Figure 1.2 Tourists visiting the active excavation site of Çatalhöyük, Turkey (Photo: Dallen J. Timothy)

identities. While the cultural artifacts of Liechtenstein and Monaco might not wield a sense of global importance, they are certainly of national importance as they help justify the existence of these improbable microstates and materialize the foundations of their national heritagescapes.

The tourism market for heritage sites has been well researched over the years (Adie & Hall, 2017; Jewell & Crotts, 2002; Timothy & Boyd, 2003). The motives for visiting archaeological and other historic localities vary widely from person to person and site to site. Many people visit for educational reasons – informal education as they seek edification and experience, and formal education, when such visits are required as part of a prescribed school curriculum. Other people visit to fulfil their curiosity about a place, person or event, while some consumers drop in to satisfy a personal interest or hobby. Timothy (2011) suggests that motivations for visiting can be seen on a spectrum. On one end are deep-seated motivations of personal interest that cause people to visit heritage sites to learn or become immersed in something beyond their normal routines. These may be referred to as serious or hard-core heritage enthusiasts, who meticulously prepare for their visits by studying and planning. On the opposite end, casual heritage consumers often visit archaeological sites to use up excess time, stop by because they happened upon an interesting locality along the route, or desire to take a 'selfie' in front of a well-known monument as part of their broader tour itinerary. Between these two extremes are various other types of tourists who may demonstrate varying degrees of interest in archaeology.

Although relatively few studies have been undertaken on the tourist demand for archaeological experiences specifically, the demand for archaeological attractions is very similar to the market for heritage sites in general (Blasco López *et al.*, 2020; Nyaupane *et al.*, 2006). The average ages of heritage visitors vary widely, depending on where they travel and the types of activities they undertake. Nonetheless, overall they tend to be middle-aged or slightly older than other tourist segments. They are generally better-educated, more affluent, stay longer in the destination and spend more money on average than other tourists do (Adie & Hall, 2017; Alzua *et al.*, 1998; Light & Prentice, 1994; Richards, 2001; Timothy, 2011).

Higher levels of education often translate into deeper desires to explore the world and experience archaeological remains as serious heritage tourists. Likewise, higher-than-average incomes facilitate archaeology enthusiasts to travel more frequently to exotic locations, in many cases, 'collecting' archaeological sites. While many heritage tourists visit sites that are somehow connected to them personally (Poria *et al.*, 2006; Timothy, 1997), such as monuments that commemorate a battle one's grandparent might have participated in, a farmstead where an ancestor farmed, or a familial village in a diasporic homeland, it is unlikely that ancient archaeological sites would be considered personal heritage among tourists today.

Tourism and archaeology: A natural symbiosis?

While many of the earliest 19th-century excavations were funded by private institutions or wealthy individuals, oftentimes so that they could accrue artifacts for their own collections and galleries, archaeological funding later fell under the primary domain of public institutions, national or regional governments. However, during the past half century, like many other public funding priorities, archaeology has suffered from government austerity measures, so that archaeological work is now funded largely by non-profit organizations/NGOs and membership societies.

Tourism has now become the standard operating procedure for many archaeological projects as it provides symbiotic economic benefits. For archaeology, tourism covers much of the cost of continued excavations and the protection of cultural resources (Ramsey & Everitt, 2008). In fact, nowadays tourism is frequently singled out as one of the primary justifications for digs, building analysis, interpretive programs, conservation efforts and public archaeology outreach. Entrance fees into museums and archaeological parks supplement many excavations and research projects throughout the world. In some localities, entrance tickets and visitor donations are the sole source of revenue that keeps the excavations in operation (Helmy & Cooper, 2002).

For tourism, beyond its scientific, educational and conservation value, the archaeological record also wields considerable economic value (Burtenshaw, 2015; Gould & Burtenshaw, 2014; Kinghorn & Willis, 2008). As previously noted, archaeological remains are among the most visited heritage attractions in the world and are an enormous engine for economic development (Giraudo & Porter, 2010). Babalola and Ajekigbe (2007) even suggest that archaeology-based tourism is a form of pro-poor tourism – that which can benefit all segments of society, including the impoverished. While archaeology's tourism value is obvious, less apparent are its socioeconomic values, including promoting resident well-being by providing recreational and volunteer opportunities, community buy-in and civic pride, and artifacts make localities more attractive for potential new residents and outside business investors.

Tourism growth and niche market development

Many archaeologists and other heritage stewards have traditionally scorned the idea of mass tourism, because tourism can be a destructive force and is sometimes seen as antithetical to the scientific discovery and conservation roles of archaeologists, cultural resource managers and curators (Burtenshaw, 2014; Deacon, 2006). However, there is a growing realization that tourism is justifiable as a funding source and a means of educating the public about the cultural past. Archaeology-based tourism is increasingly being recognized as a manifestation of public archaeology (Lenik, 2013; Newell, 2008), so despite some heritage managers' initial reluctance to become involved in tourism, many now see it as a necessity to ensure operational longevity.

Tourism has grown significantly and exponentially since the Second World War as transportation technology improved, borders became more open, families became more affluent, education levels increased, and the world in general became a smaller place. In 1950, approximately 25 million international journeys were taken. By 1990, the number of international trips had increased to 457 million, and in 2000, 698 million international arrivals were recorded. The year 2013 surpassed the 1 billion mark, and in 2017, 1.323 billion international trips were estimated to have occurred. Tourism has grown at a steady rate of 4–6% per annum, and it is forecasted to continue growing as more destinations open up to tourism and as more people are able to travel.

Much of the growth in tourism in general, and heritage tourism specifically, can be attributed to massive marketing efforts by destinations, promotional intermediaries, government agencies, and individual business owners. While there are ways in which marketing can effectively support the sustainable use of archaeological resources (Chhabra, 2010), most global destinations have adopted a blind promotion approach (boosterism) in which increasingly higher numbers of visitors are the ultimate goal through place branding.

As mentioned earlier in this chapter, archaeological sites and other heritage remains have been the focus of massive branding efforts for decades, which has created iconic images of tourist destinations associated with certain heritage artifacts. Some countries have their own heritage brands. For instance, in the United States, the US National Park Service maintains the National Register of Historic Places and the National Historic Landmarks program, both of which designate special places throughout the country as being particularly meaningful to the historic American identity, and may have a tourism value. At the international level, the most obvious brand is the UNESCO World Heritage Site identifier (Poria *et al.*, 2011). Almost every country on earth has one or more World Heritage Sites (WHS). Several countries are clambering to inscribe as many of their historic localities as possible on the list, under the assumption that the UNESCO brand will somehow increase tourist visitation (Chih-Hai *et al.*, 2010; Vargas, 2018), multiply government or international funding, and expand a region's cultural sophistication and national pride and identity on the world stage (Jimura, 2011; Tarawneh & Wray, 2017). Critics of the WHS brand have noted the overly political nature of UNESCO's inscription process and the concomitant favoritism, nepotism, and prejudices associated with it (Meskell, 2015; Vargas, 2018). In fact, according to Adie *et al.* (2018: 399), WHS designation's 'importance may be tied more to political interests than economic advancement'.

Several studies have concluded that WHS listing does not guarantee increased visitation. Most studies show a considerable mix of results in questioning whether or not the WHS trademark enhances arrivals (Adie et al., 2018; Buckley, 2004; Hall & Piggin, 2001; Huang et al., 2012).

One characteristic of 21st century tourism is the growth of niche tourisms, or at least the recent recognition of niche markets that might have already existed. Archaeotourism, as noted earlier, is recognized as a unique form of heritage tourism wherein the goal of traveling is to visit places of archaeological significance and to learn about the cultural heritage of places through excavations, displays and interpretive programs. Archaeology-based volunteer tourism is another important niche product that involves people paying their own travel costs and program fees, and donating time and energy to participate in archaeology fieldwork. Their motives may be altruistic, such as conservation mindedness and a desire to help the communities where the digs are located, or they may be more self-oriented, such as earning course credits, practicing a language, or experiencing a unique tourist destination.

A third niche market is religious tourism, which includes both pilgrims and non-pilgrim tourists. Religious tourists are prodigious consumers of archaeology. Many of the shrines and buildings they visit and worship in, or the relics they desire to see, are of ancient origin and built upon the ruins of previous historic structures. Much religious travel also venerates archaeological ruins that were once important holy places (e.g. the Cathedral of St Andrew, Scotland) or shrines that have been revered and continuously inhabited since ancient times (e.g. the Church of the Holy Sepulchre in Jerusalem).

Even the broad notion of sport tourism may have elements of archaeology, particularly when enthusiasts visit the cultural hearths of certain games, or early stadia and arenas, such as the famous handball courts of the Maya civilization in Mesoamerica (Magnoni et al., 2007) (Figure 1.3). Agritourism is another special interest form of tourism that involves visiting farms, participating in food production, and enjoying agricultural landscapes, and while this type of tourism is not commonly associated with archaeology, it sometimes is. For instance, the ancient rice terraces of East and Southeast Asia and the agricultural systems that formed them are part of an ancient agrarian system that continues to link the past with the present and has become a focal tourist attraction in places such as China, the Philippines and Indonesia (Sun et al., 2010, 2011).

Likewise, spa tourism has existed for centuries and became particularly popular during the Roman Empire in locations throughout Europe and the Middle East. While many ancient spa ruins have been excavated and function as generic heritage attractions, there remain several important spa destinations that have been in use since ancient times. For example, the thermal baths in Bath, England, which the Romans developed and frequented, were revived by the British aristocracy in the 17th century and

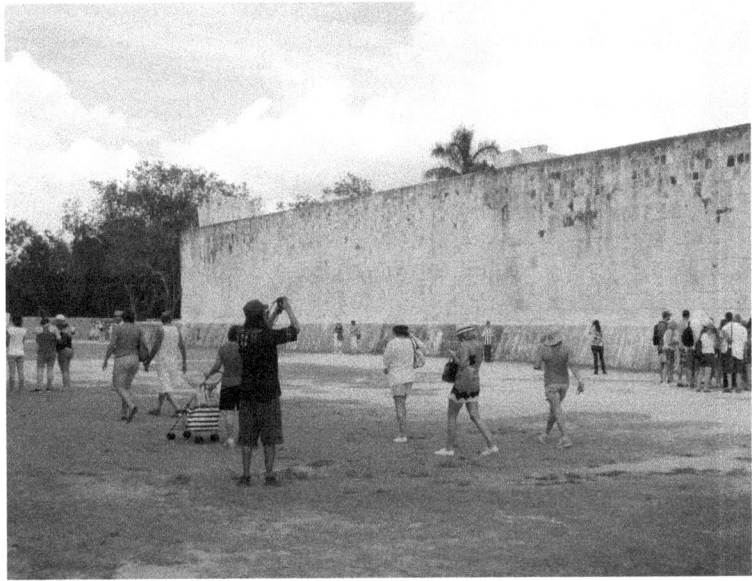

Figure 1.3 Handball courts at the ancient Maya city of Chichen Itza provide archaeological evidence of the ancient development of sport (Photo: Dallen J. Timothy)

have remained popular since that time (Murphy, 2012), although visitors are no longer permitted to swim in the original Roman bath due to health concerns. The city's thermal waters have been diverted to newer baths. Similarly, the therapeutic hot springs of Spa, Belgium, have been used continuously since the 14th century.

While many scholars have argued that niche types of tourism, special-interest tourism, and alternative forms of tourism (e.g. Agarwal et al., 2018; Novelli, 2005; Weaver, 2006) exhibit fewer of the negative impacts commonly attributed to mass tourism because they have smaller markets that are more narrowly focused, and are more sensitive in their behaviors, what we are now beginning to see is the growth of mass alternative tourism or mass special-interest tourism. This 'massification' of niche and special-interest tourisms generates the same problems and issues that face traditional leisure-oriented travel, including resource destruction.

The destruction of archaeology: Tourism and physical development

Tourism is widely acknowledged as a positive force from an economic development perspective. As previously noted, it brings in tax revenues, stimulates entrepreneurial activity and provides employment for destination residents. It can also help justify conservation and interpretive programs at archaeological sites. However, there is a distinct downside to tourism almost everywhere it occurs. Tourism brings in its wake many negative social, cultural and environmental challenges, which are exacerbated and magnified when tourism is allowed to grow spontaneously, without careful planning (Comer, 2014; Timothy, 1994, 1999).

Post-World War Two tourism grew organically in most cases, bringing with it discontent and discord in many destination communities, where residents began to despise outsiders and what they represented: disrespect, prostitution, drugs, crime and crowdedness. As well, tourists' demand for tangible artifacts and intangible culture caused living heritage to be altered to meet the needs of the visitors. Neocolonialist relationships underscored by socioeconomic inequity, advantage-taking and thuggery became the norm in many destinations, and many places became too reliant on tourism for their economic well-being, which is particularly problematic among small states that have few other development options.

Together with these socioeconomic and cultural challenges, mass tourism also caused the deterioration of natural and built environments, permanently affecting certain species of flora and fauna, and deteriorating the material culture substantially through graffiti, vandalism, excess rubbish, and physical wear and tear (Timothy, 1994, 2011; Timothy & Nyaupane, 2009a). It should be noted, however, that ancient graffiti before the advent of mass tourism, is now recognized as part of the valued heritagescapes of many archaeological sites and monuments (Figure 1.4). At Luxor, Egypt, excessive visitation has caused increased moisture in the air, which has faded colors in some of the reliefs, and tourists climbing the pyramids, urinating on them, and entering structures that were marked off limits have damaged the ancient structures (Enseñat-Soberanis *et al.*, 2019). In ancient Petra, Jordan, masses of tourists walking on and touching delicate sandstone surfaces have severely damaged its sculptures and monuments (Comer, 2012; Mustafa & Abu Tayeh, 2011; Tarawneh & Wray, 2017) and the explosive growth of tourism in Cambodia since the 1990s has brought about many negative impacts on the temples of Angkor Wat (Winter, 2008).

While tourism is often faulted for its destructive characteristics, some scholars acknowledge that the industry also plays a role in conserving and protecting the past (Hoffman *et al.*, 2002). Earnings from tourism, as discussed previously in this chapter, not only help prolong the archaeological inquiry in a specific locality, they can also be utilized to effect conservation and restoration programs, including the establishment of archaeological parks and museums.

Clearly, tourism is not the only culprit of the destruction of the archaeological record. Agriculture, heavy industry and traffic pollution, development projects, religious fanaticism and war, and looting are even more destructive to cultural artifacts and the historic record than tourism is. Clearing land for agricultural purposes, tilling soil and applying fertilizers

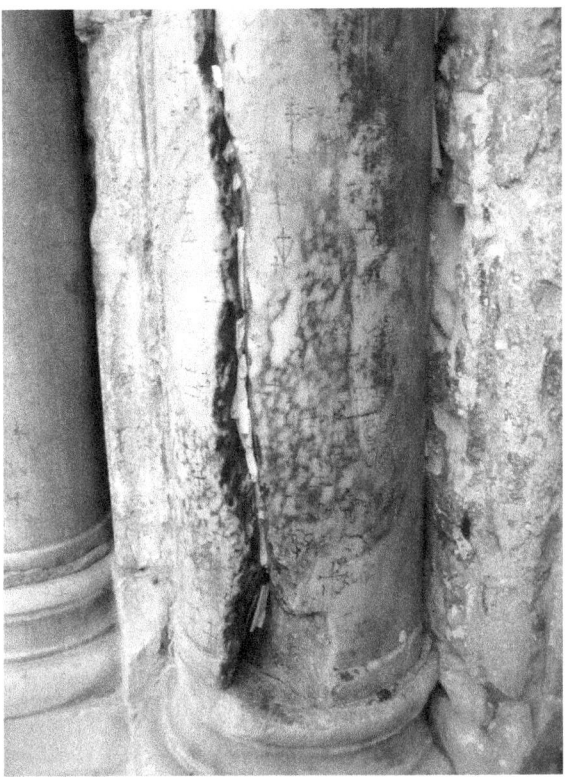

Figure 1.4 The graffiti left behind by Medieval pilgrims in the Holy Land has become part of the heritage appeal of some attractions (Photo: Dallen J. Timothy)

and pesticides all have impacted the archaeological record (Navazo & Díez, 2008). Mining has been known to destroy archaeological remains, and the airborne and waterborne toxins from heavy industry and air pollutants from heavy vehicle traffic discolor historic structures and deteriorate the physical integrity of ancient monuments and cultural remains (Kuzmichev & Loboyko, 2016). One of the best documented instances of this on an ancient monument is that of the Taj Mahal in India, which has experienced considerable decay in recent decades (Gauri & Holdren, 1981; Pandey & Kumar, 2015). Road construction and the development of other infrastructure also has a poor record of damaging the archaeological record. While most developed countries today require impact assessments for large development projects, some regions remain without adequate legislation or choose not to enforce existing impact assessment laws for fear that such actions will add significant time and cost to construction projects.

In recent years, the strong link between religious fanaticism and the destruction of cultural property has been at the forefront of archaeology and heritage resource management discussions. The 2001 destruction of the ancient Buddha statues in the Bamyan Valley, Afghanistan, by Taliban rebels seems to have precipitated this phenomenon in the 21st century (Ashworth & van der Aa, 2002). Religious fanaticism's impacts on archaeology has been especially poignant since 2014 with the rise of ISIS in Syria and Iraq, and the terror organization's destruction of ancient heritage under the fictitious claim of false gods and idol worship (Turku, 2018). Relatedly, war itself is known to destroy the archaeological record of places, especially when artifacts become targets of annihilation for their national or cultural identity value.

Another salient concern is looting. Pillaging archaeological sites and looting cultural artifacts has been a problem for centuries and derives primarily from economic motivations as diggers loot sites to sell artifacts to intermediaries and collectors. People travel to collect or purchase ancient artifacts, or to deal in them. This has led to the widespread pillaging of archaeological sites throughout the world. When tourism provides a marketplace for the illicit trade in ancient artifacts, there will always be suppliers who are willing to dig in archaeologically sensitive areas.

While destruction by farming, heavy industry and traffic, infrastructure development, religious fanaticism, war, and looting might appear disconnected from tourism, it is far from being disconnected. In fact, there are very clear connections between tourism and these other forces. For example, in heavily touristed areas, increased food production is required to meet the needs of tourists' alimentary demands. A vibrant tourism industry increases vehicle traffic considerably; the need for access to destinations and attractions accelerates road building; a growing tourism sector requires additional hotels, resorts and restaurants; war and religious fanaticism have been known to be funded, in part at least, by collectors' (including tourists) expenditures on looted artifacts (Mustafa, 2019), and looters sell their spoils to unsuspecting leisure tourists, serious antiquities collectors and unscrupulous middlemen (Di Lernia, 2005).

Given tourism's negative impacts on historic environments, as well as the adverse effects of war, anti-heritage extremism, farming and physical development, the need for archaeological protection has never been more absolute. The growth of public archaeology has helped alleviate some of these concerns, but much more work is required. The caretakers of archaeological heritage work hand in hand with governments to enact protective legislation, develop interpretive programs, and establish effective site management plans.

Enseñat-Soberanis and his colleagues (2019) analyzed the management strategies of 11 well-known archaeological sites in Europe, the Middle East, China and Latin America. They concluded that the most common approaches to mitigating the negative impacts of tourism on tangible heritage are threefold: restrictive strategies, redistributive strategies and interpretive strategies. Restrictive policies include limiting visitors' ability to touch or make physical contact with relics and also to limit the number of people who can visit at one time or during one period, by establishing carrying capacities. Secondly, common redistributive strategies include dispersing visitors through time and space. This entails, allocating groups to less busy times and perhaps enacting quotas on visitor numbers during peak periods, as well as allowing visitors or certain group sizes to access only certain areas of a site. Third, interpretive strategies aim to educate visitors by communicating the importance of the heritage value of the site and persuading visitors to change their behaviors (Enseñat-Soberanis *et al.*, 2019). The results of their study reflect findings similarly to those of many other studies over the years that have examined how best to manage visitors in delicate archaeological areas (Timothy, 2011).

Devising innovative conservation and heritage management tools is critical in today's fast-paced consumer and technology-driven society. Traffic control and visitor flow and congestion management are key in protecting resources and providing satisfying consumer experiences. Other common means of managing heritage and its visitors include limiting contact with artifacts, pricing policies, providing high-quality experiences that will encourage visitors to be more respectful, utilizing principles of sustainability in promotion efforts, and providing entertaining, engaging, and informative interpretive programs (Timothy & Boyd, 2003).

Politics of the past

The very concept of heritage is extremely partisan and contested. Archaeological heritage is frequently at the forefront of the politicization of the past as governments or agencies maneuver heritage to achieve a desired result and exercise authority over places, people and processes. Authorities manipulate tourism in many different ways (e.g. embargoes, travel warnings, visa restrictions and siding with allies), some of which are directly related to heritage. One of the most obvious is the use of heritage to foster domestic patriotism, national solidarity and a heroic state narrative (Timothy, 2007).

In this sense, the archaeological record is employed to authenticate state territorial claims, legitimize governments in power, venerate national heroes, idealize the homeland, empower certain population cohorts while simultaneously disempowering others, and corroborate the official textbook version of history. Similarly, archaeological heritage is commonly used as propaganda for foreign visitors to 'discredit negative events from the past, while extolling the virtues of the past and present' (Timothy & Nyaupane, 2009b: 46). In this situation, foreign tourists are encouraged to visit cultural sites that best reaffirm the nationalist chronicle and reinforce national ideals (Murakami, 2008). Likewise, during and after European colonial rule, museums and other heritage sites were programmed to reflect colonial worldviews, thereby downplaying or simplifying the importance of local heritage (Tahan, 2010b). Finally, archaeology is sometimes deployed to erase or disprove opposing views and parts of heritage that do not play into the national story, creating a sense of social amnesia (Adams, 2010; Timothy & Boyd, 2003) which, in extreme autocratic situations, can be supported by the state-engineered or state-discredited archaeological record, raising the question about 'whose archaeology is excavated and for what purpose'. According to Timothy (2007: xiii), 'Unfortunately and predictably, most victims of societal amnesia have been ethnic and racial minorities, women and other "marginal" peoples, and this has resulted in their lives and struggles being hidden from public view'.

There are many examples of this throughout the world. Tahan (2014) discusses how Lebanese museums have long favored the country's Christian history over its Muslim past because of its desire to portray a stronger western orientation. Two of the best documented examples can be found in the United States. In that country, societal amnesia and selective archaeological records long plagued the European and Native American story, where heritage and archaeology were politicized to favor the white American notion of 'manifest destiny' – the 19th-century dogma that the territorial expansion of the United States through the frontiers of North America was inevitable, justifiable, righteous, and endorsed by God. Taming the land for white habitation was part of the goal, which included 'taming' the 'savages' that occupied the land. Likewise, the powers that controlled the national narrative for centuries also chose to de-emphasize the archaeology of African American slaves far into the 20th century to favor the white American storyline (Singleton, 2016). Similar conditions were perpetrated by European metropolitan powers in the Asian, African, Pacific, Caribbean and Latin American settler societies they created through colonialism.

While tourism has played a role in perpetuating these biased narratives, proving its role in the disempowerment of native peoples and ethnic minorities, tourism that is based on truer and more balanced renditions of indigenous archaeology has the potential to empower native peoples socioeconomically and politically (Vargas, 2018). Only in the latter part of the 20th century and into the new millennium have we seen this condition improve and demoralized communities become increasingly empowered through more objective archaeological interpretations and tourism (Gallivan, 2011; Meskell & Preucel, 2004; Parks, 2010; Singleton, 2016). These issues appear to be particularly poignant in descendant communities who feel they have legitimate claims to ownership of local ruins (Nyaupane et al., 2006; Pacifico & Vogel, 2012; Parks, 2010).

Part of heritage politics lies in archaeological interpretation. In the context of heritage, interpretation means to tell the story, to reveal the significance of the place or site. Although interpretation is typically discussed in the context of cultural resource management, owing to its value in providing information, controlling crowds, eliciting better tourist behavior, and offering safety and protective warnings, interpretation has also been the focus of much debate on the politics of archaeology (Li & Qian, 2017; Timothy, 2011; Timothy & Boyd, 2003).

There has been a lot of research on the interpretive responsibilities of tour guides at cultural sites and their role as brokers of knowledge (e.g. Ababneh, 2018; Weiler & Black, 2015). Their position wields considerable power in disseminating knowledge to tourist consumers, as they explain events, people and places according to what they want, or have been trained, to provide to visitors (Ababneh, 2018; Zhao & Timothy, 2017).

The Focus of this Book

This collection of essays aims to provide a conceptually sound overview of many issues confronting the interface of archaeology and tourism. While this crossover between tourism and archaeology is extremely diverse and could encompass many volumes, we have managed to examine several main themes that are particularly relevant today.

The following three chapters examine the symbiotic relationships between tourism and archaeology. In Chapter 2, Laurence Gillot describes how archaeologists have not always got along with tourism and its promoters, although these relationships are on the mend as archaeologists realize the need to work with, rather than against, tourism. This sometimes antagonistic relationship stems from the direct and indirect damage that frequently occurs to cultural artifacts and historic environments through mass tourism. Gillot suggests that nowadays, archaeologists are more willing to play a larger advocacy role in tourism because they see the need for the visitor industry – a realization that has changed the relationships from one of confrontation to one of collaboration. In the next chapter, Paul Burtenshaw continues to explore the mutually beneficial relationships between tourism and archaeology, particularly from and economic values, or economic capital, perspective. Owing to its socioeconomic value, tourism's use of archaeology can foster symbiotic relationships in a way that generates employment and regional income in the destination. Tourism, Burtenshaw contends, is one important way archaeology can give back to the community to which it belongs. Site and destination planning must be carefully considered as tourism is promoted and continues to grow. In her chapter 'Privatization, Archaeology and Tourism', Işilay Gürsu ponders the role of privatization in the heritage/archaeology sector. By transforming ownership from the public domain to the private sector, state goods and services become private goods and services, which can increase efficiency and save state budgets for other public needs. Tourism has an important role to play in archaeological privatization, although there is still much resistance among heritage stewards, some government officials and some community stakeholders against private ownership of archaeology.

The focus of the next three chapters is the growth of archaeology-based tourism through marketing efforts and the identification of niche forms of archaeotourism, namely volunteer archaeology vacations and religious tourism. Chapter 5, by Alan Fyall, Anna Leask and Sarah Barber, provides a fascinating overview of marketing in the realm of heritage/archaeology tourism. Using empirical evidence from Mexico and a marketing campaign in Scotland, they argue that marketing archaeological heritage can be effective in distinguishing heritage destinations from their competitors. In this sense, archaeological remains provide a competitive advantage over other would-be competitor destinations. Fyall, Leask and Barber discuss other critical marketing principles, such as product bundling, through the creation of archaeological routes, which strengthens a region's heritage product even more. Branding, especially the World Heritage Site brand, is increasingly being used as a mechanism to build awareness, entice people to visit and once again to create a competitive advantage over other attractions and destinations. The authors also acknowledge the increasingly important role of virtual reality, augmented reality and other digital tools in creating place images and being more competitive in the tourism marketplace.

The focus of Chapter 6 (Dallen Timothy) is archaeology-based volunteer tourism. While volunteer tourism is a growing niche market, particularly in the developing world, we know relatively little about archaeology volunteering as a form of volunteer tourism. Timothy reviews the motives of volunteering in archaeological settings and provides an overview of the market, suggesting that many people pursue this work activity during their leisure time either as a personal, self-oriented pursuit or for more altruistic purposes that aim to discover and provide enjoyable heritage experiences for visitors. He outlines three examples of reasons people desire to volunteer at archaeological excavations or in related tasks – to further their own hobby interests, religious devotion, and an academic fascination with the place, time or culture under study. There is a vast network of intermediaries, agencies, promoters and scientists that all work together to facilitate archaeology-based volunteer tourism as a growing commercial enterprise.

Like the two before it, the focus of Chapter 7 by Nour Farra-Haddad, is understanding an increasingly important tourism niche – religious tourism, or pilgrimage – and its interdependence with religious archaeology. Archaeology is an important medium between religious tourists and the sacred sites they visit. Utilizing empirical evidence from Lebanon and various other countries, Farra-Haddad illustrates how some places are believed to be divinely appointed as sacred, regardless of who visits and what religion controls the sacred space at any given time. Religious site stratigraphy supports this conclusion, as many religious structures have been built one upon the other over centuries or millennia. From this, she identifies a religious archaeology lifecycle – discovery, acceptance, veneration, decline and disappearance – which may also include a revivification of the locale by a different faith later in history. Religious archaeology reflects how places were sanctified in the past and how their religious geography remains in the present.

The third group of chapters emphasizes the overgrowth of tourism and how it has damaged the built environment and caused the consumption of archaeological remains. They also provide an understanding about the need to protect these resources for and from tourism. In Chapter 8, Lina Tahan summarizes many forces that contribute to the direct and indirect destruction of the archaeological record. She considers the salient role of urban development, agriculture, road building, natural disasters, religious fanaticism, and mass tourism as destroyers of the tangible human patrimony. Modern-day mass tourism has become one of the most destructive forces against cultural artifacts through direct contact, wear and tear, vandalism and physical development. Good planning, pro-heritage policies and careful implementation are necessary to protect the archaeological record.

Continuing with the notion of damage, Dallen Timothy's chapter (Chapter 9) on the illicit trade in antiquities and cultural artifacts spotlights the problem of looting and the illicit antiquities trade. Much of this problem began with the world's exploration periods during the Middle Ages and throughout the colonial era, as the elites, including many museums in the European metropoles, vied for the treasures of their faraway colonies. Collectors who are willing to pay high prices for valuable relics continue to fuel the trade in illegal antiquities, and tourism is part of the problem. Tourism feeds looting in a variety of ways, including peddlers selling found items to tourists, licensed retailers selling to tourists, licensed dealers hawking illegal artifacts, brokers traveling to buy and sell, and tourists digging or finding artifacts themselves. These activities can also result in fake artifacts and setting tourists up to be ripped off. Timothy also examines the critical role of tourism's unlawful consumption of cultural artifacts contributing to the activities of terrorist organizations in various unstable countries, where archaeological sites are routinely pillaged and the loot sold as tourist consumer goods.

The destruction of the tangible past as described in Tahan and Timothy's chapters, leads Jennifer Mathews to write about 'protecting the archaeological past in the face of tourism demand'. In Chapter 10, she articulates how tourists frequently have a shallow sightseeing experience; their appreciation of heritage is usually tangential and is part of the global phenomenon of mass tourism. Mathews considers community engagement as an important tool for protecting the built environment, in tandem with an increased appreciation of a more ethical treatment of archaeological remains. This line of thinking will help balance the needs of tourism, the community's economic needs, and heritage site protection. By becoming more involved in the management and sustainable marketing of heritage sites, archaeologists can help protect the record they seek to learn from and conserve.

The final two content chapters deal specifically with many political aspects of archaeology and tourism. In Chapter 11, Sue Hodges eruditely scrutinizes the role of interpretation in educating the public and inducing action on the part of visitors and communities whose heritage is only display. Interpretation has a long history, but it has continued to evolve, and it is extremely complex. Today, not only is interpretation and its various media an important tool for learning, teaching and enjoying, it is also a highly political instrument that is manipulated by people in power to create the narrative they wish to convey. In fact, it is one of the most political elements of the archaeology and tourism stage. Archaeological interpretation has been used to uphold disputed claims for authority and authenticity, and it nearly always has multiple meanings that prescribe heritage to one group and proscribe it to another, thereby fortifying one group's claims over another. This has been a problem in the past with regard to indigenous and colonial peoples and slaves and slave masters, for example.

Gai Jorayev (Chapter 12) also deconstructs the political inner lining of archaeology and tourism in the context of nationalism. He describes the frequent political manipulation of archaeology for the purpose of nation building and examines the state's role in marketing, branding, interpreting and conserving the archaeological record. Archaeology is often treated as a conduit for advancing the ideologies of the state, such as territorial claims, boundary changes, ethnic identity and indigeneity, or racial segregation. UNESCO's World Heritage Site inscription, according to Jorayev, is a driver of nationalism and tourism, so that what was previously national heritage becomes universal heritage through this inscription process.

In conclusion, the relationships between archaeology and tourism are heterogeneous, complex, and challenging. That archaeological remains are among the most visited heritage attractions in the world is without question, and they are frequently used by the state and its tourism machinery to brand itself and create an advantage over its market competitors. Despite the world's political, economic and security vicissitudes, tourism continues to grow unabated. Archaeologists have now come to terms with the idea that they too must be involved in tourism from at least two contemporaneous perspectives. First, tourism provides a social and economic justification for archaeologists' scientific explorations, and second, their skillsets in managing material culture are extremely important in protecting the built environment from an industry that has the potential to destroy the very assets upon which it is based. These evolving perspectives have manifested in traditional archaeology having expanded beyond dig sites to include cultural resource management training that enlarges archaeologists' and other heritage stewards' role into the sphere of management and the visitor interface. Tourism is, in a sense, the ultimate form of public archaeology.

Beyond its tourism-specific challenges, archaeology and tourism is a highly political relationship that can simultaneously empower communities or disempower them. Governors determine what histories will be told, and interpreters function as the on-site storytellers who perpetuate certain myths or truisms that lend a touristic intrigue to heritage localities. There has been an obvious pattern of archaeological and heritage manipulation to meet the needs of the people in power.

As the chapters in this book make abundantly clear, there is growing research interest in issues surrounding the juxtaposition of tourism and archaeology. The exponential appearance of articles that meld the two themes in journals such as *Annals of Tourism Research*, *Journal of Heritage Tourism*, *Tourism Management*, *International Journal of Heritage Studies*, *Public Archaeology* and *World Archaeology*, attests to the growing interest in archaeology among tourism scholars and interest in tourism among archaeologists. The aim of this book is to complement, rather than replace, the excellent work already done by Walker and Carr (2013), Gould and Pyburn (2017) and others in their analyses of the unique relationships between archaeology and tourism. Despite their efforts and ours, we have a long way to go before the multifarious relationships between the two phenomena are fully understood, if that is even possible.

References

Ababneh, A. (2018) Tour guides and heritage interpretation: Guides' interpretation of the past at the archaeological site of Jarash, Jordan. *Journal of Heritage Tourism* 13 (3), 257–272.

Adams, J.L. (2010) Interrogating the equity principle: The rhetoric and reality of management planning for sustainable archaeological heritage tourism. *Journal of Heritage Tourism* 5 (2), 103–123.

Adie, B.A. and Hall, C.M. (2017) Who visits World Heritage? A comparative analysis of three cultural sites. *Journal of Heritage Tourism* 12 (1), 67–80.

Adie, B.A., Hall, C.M. and Prayag, G. (2018) World Heritage as a placebo brand: A comparative analysis of three sites and marketing implications. *Journal of Sustainable Tourism* 26 (3), 399–415.

Agarwal, S., Busby, G. and Huang, R. (eds) (2018) *Special Interest Tourism: Concepts, Contexts and Cases*. Wallingford: CABI.

Akagawa, N. and Smith, L. (eds) (2019) *Safeguarding Intangible Heritage: Practices and Politics*. London: Routledge.

Alzua, A., O'Leary, J.T. and Morrison, A.M. (1998) Cultural and heritage tourism: Identifying niches for international travelers. *Journal of Tourism Studies* 9 (2), 2–13.

Ashmore, W. and Sharer, R.J. (2014) *Discovering Our Past: A Brief Introduction to Archaeology* (6th edn). New York: McGraw-Hill.

Ashworth, G.J. and van der Aa, B.J. (2002) Bamyan: Whose heritage was it and what should we do about it? *Current Issues in Tourism* 4 (5), 447–457.

Babalola, A.B. and Ajekigbe, P.G. (2007) Poverty alleviation in Nigeria: Need for the development of archaeo-tourism. *Anatolia* 18 (2), 223–242.

Blasco López, M.F., Recuero Virto, N., Aldas Manzanob, J. and García-Madariaga, J. (2020) Archaeological tourism: Looking for visitor loyalty drivers. *Journal of Heritage Tourism* 15 (1), 60–75.

Buckley, R. (2004) The effects of World Heritage listing on tourism to Australian national parks. *Journal of Sustainable Tourism* 12 (1), 70–84.

Burtenshaw, P. (2014) Mind the gap: Cultural and economic values in archaeology. *Public Archaeology* 13 (1–3), 48–58.

Burtenshaw, P. (2015) *Archaeology and Economic Development*. London: Routledge.

Carman, J. (2009) Where the value lies: The importance of materiality to the immaterial aspects of heritage. In E. Waterton and L. Smith (eds) *Taking Archaeology Out of Heritage* (pp. 192–208). Newcastle-upon-Tyne: Cambridge Scholars.

Casson, L. (1994) *Travel in the Ancient World*. Baltimore: Johns Hopkins University Press.

Chhabra, D. (2010) *Sustainable Marketing of Cultural and Heritage Tourism*. London: Routledge.

Chi-Hai, Y., Hui-Lin, L. and Chia-Chun, H. (2010) Analysis of international tourist arrivals in China: The role of World Heritage Sites. *Tourism Management* 31 (6), 827–837.

Comer, D.C. (2012) *Tourism and Archaeological Heritage Management at Petra: Driver to Development or Destruction?* Baltimore: Springer.

Comer, D.C. (2014) Threats to the archaeological heritage in the *laissez-faire* world of tourism: The need for global standards as a global public good. *Public Archaeology* 13 (1–3), 123–134.

Davidson, J.M. and Brandon, J.C. (2012) Descendent community partnering, the politics of time, and the logistics of reality: Tales from North American, African diaspora, archaeology. In R. Skeates, C. McDavid and J. Carman (eds) *The Oxford Handbook of Public Archaeology* (pp. 605–628). Oxford: Oxford University Press.

Deacon, J. (2006) Rock art conservation and tourism. *Journal of Archaeological Method and Theory* 13 (4), 376–396.

Di Lernia, S. (2005) Incoming tourism, outgoing culture: Tourism, development and cultural heritage in the Libyan Sahara. *The Journal of North African Studies* 10 (3/4), 441–457.

Emerick, K. (2009) Archaeology and the cultural heritage management 'toolkit': The example of a community heritage project at Cawood, North Yorkshire. In E. Waterton and L. Smith (eds) *Taking Archaeology Out of Heritage* (pp. 91–116). Newcastle-upon-Tyne: Cambridge Scholars.

Enseñat-Soberanis, F., Frausto-Martínez, O. and Gándara-Vásquez, M. (2019) A visitor flow management process for touristified archaeological sites. *Journal of Heritage Tourism* 14 (4), 340–357.

Fouseki, K. (2009) I own, therefore I am: Conflating archaeology with heritage in Greece: A possessive individualist approach. In E. Waterton and L. Smith (eds) *Taking Archaeology Out of Heritage* (pp. 49–65). Newcastle-upon-Tyne: Cambridge Scholars.

Gallivan, M. (2011) The archaeology of native societies in the Chesapeake: New investigations and interpretations. *Journal of Archaeological Research* 19 (3), 281–325.

Gauri, K.L. and Holdren, G.C. (1981) Pollutant effects on stone monuments. *Environmental Science & Technology* 15 (4), 386–390.

Giraudo, R.F. and Porter, B.W. (2010) Archaeotourism and the crux of development. *Anthropology News* 51 (8), 7–8.

Gould, P.G. and Burtenshaw, P. (2014) Archaeology and economic development. *Public Archaeology* 13 (1–3), 3–9.

Gould, P.G. and Pyburn, K.A. (eds) (2017) *Collision or Collaboration: Archaeology Encounters Economic Development*. Cham, Switzerland: Springer.

Graham, B., Ashworth, G.J. and Tunbridge, J.E. (2000) *A Geography of Heritage: Power, Culture and Economy*. London: Arnold.

Hall, C.M. and Piggin, R. (2001) Tourism and World Heritage in OECD countries. *Tourism Recreation Research* 26 (1), 103–105.

Helmy, E. and Cooper, C. (2002) An assessment of sustainable tourism planning for the archaeological heritage: The case of Egypt. *Journal of Sustainable Tourism* 10 (6), 514–535.

Hodder, I. (2012) *Entangled: An Archaeology of the Relationships between Humans and Things*. Oxford: Wiley.

Hoffman, T.L., Kwas, M.L. and Silverman, H. (2002) Heritage tourism and public archaeology. *The SAA Archaeological Record* 2, 30–33.

Holtorf, C. (2007) *Archaeology is a Brand: The Meaning of Archaeology in Contemporary Popular Culture*. Walnut Creek, CA: Left Coast Press.

Huang, C.H., Tsaur, J.R. and Yang, C.H. (2012) Does World Heritage List really induce more tourists? Evidence from Macau. *Tourism Management* 33 (6), 1450–1457.

Jimura, T. (2011) The impact of World Heritage Site designation on local communities–A case study of Ogimachi, Shirakawa-mura, Japan. *Tourism Management* 32 (2), 288–296.

Jewell, B. and Crotts, J.C. (2002) Adding psychological value to heritage tourism experiences. *Journal of Travel & Tourism Marketing* 11 (4), 13–28.

Kinghorn, N. and Willis, K. (2008) Valuing the components of an archaeological site: An application of Choice Experiment to Vindolanda, Hadrian's Wall. *Journal of Cultural Heritage* 9 (2), 117–124.

Kuzmichev, A.A. and Loboyko, V.F. (2016) Impact of the polluted air on the appearance of buildings and architectural monuments in the area of town planning. *Procedia Engineering* 150, 2095–2101.

Lenik, S. (2013) Community engagement and heritage tourism at Geneva Estate, Dominica. *Journal of Heritage Tourism* 8 (1), 9–19.

Li, H. and Qian, Z. (2017) Archaeological heritage tourism in China: The case of the Daming Palace from tourists' perspective. *Journal of Heritage Tourism* 12 (4), 380–393.

Light, D.C. and Prentice, R.C. (1994) Who consumes the heritage product? Implications for European heritage tourism. In G.J. Ashworth and P.J. Larkham (eds) *Building a New Heritage: Tourism, Culture and Identity in the New Europe* (pp. 90–118). London: Routledge.

Magnoni, A., Ardren, T. and Hutson, S. (2007) Tourism in the Mundo Maya: Inventions and (mis) representations of Maya identities and heritage. *Archaeologies* 3 (3), 353–383.

Meskell, L. (2015) Transacting UNESCO World Heritage: Gifts and exchanges on a global stage. *Social Anthropology* 23 (1), 3–21.

Meskell, L. and Preucel, R.V. (eds) (2004) *A Companion to Social Archaeology*. Oxford: Blackwell.

Moshenska, G. (ed.) (2017) *Key Concepts in Public Archaeology*. London: University College London Press.

Murakami, D. (2008) Tourism development and propaganda in contemporary Lhasa, Tibet Autonomous Region (TAR), China. In J. Cochrane (ed.) *Asian Tourism: Growth and Change* (pp. 55–68). Amsterdam: Elsevier.

Murphy, P. (2012) Resort management analysis: Current and future directions. In C.H.C. Hsu and W.C. Gartner (eds) *The Routledge Handbook of Tourism Research* (pp. 324–338). London: Routledge.

Mustafa, M.H. (2019) Intangible heritage and cultural protection in the Middle East. In D.J. Timothy (ed.) *Routledge Handbook on Tourism in the Middle East and North Africa* (pp. 57–70). London: Routledge.

Mustafa, M.H. and Abu Tayeh, S.N. (2011) The impacts of tourism development on the archaeological site of Petra and local communities in surrounding villages. *Asian Social Science* 7 (8), 88–97.

Navazo, M. and Díez, C. (2008) Redistribution of archaeological assemblages in plowzones. *Geoarchaeology: An International Journal* 23 (3), 323–333.

Newell, G.E. (2008) Rhyming culture, heritage, and identity: The 'total site' of Teotihuacan, Mexico. *Journal of Heritage Tourism* 3 (4), 243–255.

Novelli, M. (ed.) (2005) *Niche Tourism: Contemporary Issues, Trends and Cases*. London: Routledge.

Nyaupane, G.P., White, D.D. and Budruk, M. (2006) Motive-based tourism market segmentation: An application to Native American cultural heritage sites in Arizona, USA. *Journal of Heritage Tourism* 1 (2), 81–99.

Pacifico, D. and Vogel, M. (2012) Archaeological sites, modern communities, and tourism. *Annals of Tourism Research* 39 (3), 1588–1611.

Pandey, A.K. and Kumar, V. (2015) Impact of environmental pollution on historical monuments of India: Conservation problems and remedial measures. *Journal of Applied Geochemistry* 17 (1), 50–55.

Parks, S. (2010) The collision of heritage and economy at Uxbenká, Belize. *International Journal of Heritage Studies* 16 (6), 434–448.

Poria, Y., Reichel, A. and Biran, A. (2006) Heritage site management: Motivations and expectations. *Annals of Tourism Research* 33 (1), 162–178.

Poria, Y., Reichel, A. and Cohen, R. (2011) World Heritage Site: An effective brand for an archaeological site? *Journal of Heritage Tourism* 6 (3), 197–208.

Ramsey, D. and Everitt, J. (2008) If you dig it, they will come! Archaeology heritage sites and tourism development in Belize, Central America. *Tourism Management* 29 (5), 909–916.

Richards, G. (2001) The market for cultural attractions. In G. Richards (ed.) *Cultural Attractions and European Tourism* (pp. 31–54). Wallingford: CABI.

Singleton, T.A. (2016) *The Archaeology of Slavery and Plantation Life*. London: Routledge.

Smith, L. and Akagawa, N. (eds) (2009) *Intangible Heritage*. London: Routledge.

Sun, Y., Min, Q.W., Cheng, S.K., Zhong, L.S. and Qi, X.B. (2010) Study on the tourism resource characteristics of agricultural heritage. *Tourism Tribune* 25 (10), 57–62.

Sun, Y., Jansen-Verbeke, M., Min, Q. and Cheng, S. (2011) Tourism potential of agricultural heritage systems. *Tourism Geographies* 13 (1), 112–128.

Tahan, L.G. (2010a) Challenging colonialism and nationalism in Lebanese archaeological museums. *Near Eastern Archaeology* 73 (2–3), 195–197.

Tahan, L.G. (2010b) New museological ways of seeing the world: Decolonizing archaeology in Lebanese museums. In J. Lydon and U.Z. Rizvi (eds) *Handbook of Postcolonial Archaeology* (pp. 295–302). Walnut Creek, CA: Left Coast Press.

Tahan, L.G. (2014) Challenging museum spaces: Dancing with ethnic and cultural diversity in Lebanon. In R. Daher and I. Maffi (eds) *The Politics and Practices of Cultural Heritage in the Middle East: Positioning the Material Past in Contemporary Societies* (pp. 135–148). London: I.B. Tauris.

Tarawneh, M.B. and Wray, M. (2017) Incorporating Neolithic villages at Petra, Jordan: An integrated approach to sustainable tourism. *Journal of Heritage Tourism* 12 (2), 155–171.

Timothy, D.J. (1994) Environmental impacts of heritage tourism: Physical and sociocultural perspectives. *Manusia dan Lingkungan* 2 (4), 37–49.

Timothy, D.J. (1997) Tourism and the personal heritage experience. *Annals of Tourism Research* 34 (3), 751–754.

Timothy, D.J. (1999) Built heritage, tourism and conservation in developing countries: Challenges and opportunities. *Journal of Tourism* 4, 5–17.

Timothy, D.J. (2007) Introduction. In D.J. Timothy (ed.) *The Political Nature of Cultural Heritage and Tourism* (pp. ix–xviii). Aldershot: Ashgate.

Timothy, D.J. (2011) *Cultural Heritage and Tourism: An Introduction*. Bristol: Channel View Publications.
Timothy, D.J. and Boyd, S.W. (2003) *Heritage Tourism*. Harlow: Prentice Hall.
Timothy, D.J. and Nyaupane, G.P. (2009a) Heritage tourism and its impacts. In D.J. Timothy and G.P. Nyaupane (eds) *Cultural Heritage and Tourism in the Developing World: A Regional Perspective* (pp. 56–69). London: Routledge.
Timothy, D.J. and Nyaupane, G.P. (2009b) The politics of heritage. In D.J. Timothy and G.P. Nyaupane (eds) *Cultural Heritage and Tourism in the Developing World: A Regional Perspective* (pp. 42–55). London: Routledge.
Towner, J. (1985) The Grand Tour: A key phase in the history of tourism. *Annals of Tourism Research* 12 (3), 297–333.
Turku, H. (2018) *The Destruction of Cultural Property as a Weapon of War: ISIS in Syria and Iraq*. Cham, Switzerland: Springer.
Vargas, A. (2018) The tourism and local development in World Heritage context: The case of the Mayan site of Palenque, Mexico. *International Journal of Heritage Studies* 24 (9), 984–997.
Walker, C. and Carr, N. (eds) (2013) *Tourism and Archaeology: Sustainable Meeting Grounds*. Walnut Creek, CA: Left Coast Press.
Waterton, E. and Smith, L. (2009) There is no such *thing* as heritage. In E. Waterton and L. Smith (eds) *Taking Archaeology Out of Heritage* (pp. 10–27). Newcastle-upon-Tyne: Cambridge Scholars.
Watson, S. (2009) Archaeology, visuality and the negotiation of heritage. In E. Waterton and L. Smith (eds) *Taking Archaeology Out of Heritage* (pp. 28–47). Newcastle-upon-Tyne: Cambridge Scholars.
Weaver, D. (2006) *Sustainable Tourism*. Oxford: Butterworth Heinemann.
Weiler, B. and Black, R. (2015) *Tour Guiding Research: Insights, Issues and Implications*. Bristol: Channel View Publications.
Wilson, R. (2009) Archaeology: Quiet on the Western Front. In E. Waterton and L. Smith (eds) *Taking Archaeology Out of Heritage* (pp. 72–90). Newcastle-upon-Tyne: Cambridge Scholars.
Winter, T. (2008) Post-conflict heritage and tourism in Cambodia: The burden of Angkor. *International Journal of Heritage Studies* 14 (6), 524–539.
Wurz, S. and van der Merwe, J.H. (2005) Gauging site sensitivity for sustainable archaeo-tourism in the Western Cape Province of South Africa. *South African Archaeological Bulletin* 60 (181), 10–19.
Zhao, S. and Timothy, D.J. (2017) The dynamics of guiding and interpreting in red tourism. *International Journal of Tourism Cities* 3 (3), 243–259.

2 Archaeologists and Tourism: Symbiosis or Contestation?

Laurence Gillot

Introduction

Archaeology and tourism have long maintained close but ambivalent relations. Both indeed share a common history, in particular from the 19th century. While archaeology emerged as a science, at the same time it offered a set of resources to the emerging tourism industry (Baram, 2008; Walker & Carr, 2013). The relations between archaeologists and tourists were then rather cordial, but as tourism continued to attract a constantly larger audience, the relationship deteriorated. The second half of the 20th century thus witnessed a widening gap between these two worlds. Even though it is still often thought, particularly in the archaeological community, that tourism and development are antithetical to the goals of heritage and archaeology, the strong emergence of conservation and sustainability philosophies in tourism during the last two decades of the 20th century made it possible to reconsider the relationship between archaeology and tourism in a more positive light. After all, tourism and archaeology may have the potential to foster mutual understanding between different stakeholders, encourage contact between cultures, and even create jobs and encourage local or national economic activities. Moreover, archaeological and tourist activities, together, have multiple and contrasted impacts on destination environments, both positive and negative, depending on the point of view and the scale of analysis.

This chapter aims to analyse the ambiguous relationships between archaeology and tourism, and archaeologists and tourists, while providing an overview of current issues and key concepts in tourism and in the work of archaeologists. In the first part, the chapter considers the bulk and characteristics of archaeological resources used in heritage tourism, with regard to an increase in public interest in access to these resources. Regarding tourism's use of archaeological work and resources, the chapter then considers the relationship between tourism and archaeology.

In particular, the chapter provides an overview of the literature relating to the archaeological debate with regard to the current history of the discipline. Finally, this work bridges the gap between archaeology and tourism by underlining the potential benefits of an applied archaeology, devoted to archaeological resource management and public presentation.

Archaeological Resources for Tourism

The practice of archaeology has some potential to generate increased public appreciation for the accounts of the past, and public access to interpreted sites and archaeological collections have long been recognized as a subject of tourism development (Timothy, 2011). Archaeological resources provide visitors an opportunity to view visible remains in their original context, see collections in museum exhibits, and read, hear and see interpretations of what has been learned (Melotti, 2011; Walker & Carr, 2013). Archaeology is consequently a major component of the tourism industry in both developed and developing countries. In this context, the concepts of 'archaeotourism' and 'archaeological tourism' can evoke a form of cultural tourism, which aims to promote public interest in archaeology and the conservation of historical assets (Giraudo & Porter, 2010; Wurz & van der Merwe, 2005). Archaeological tourism thus includes all products associated with the promotion of public archaeology, including visits to archaeological sites, museums, interpretation centres and archaeological parks, festivals, or theatres. However, while some archaeological resources may be appropriate for public consumption through heritage tourism and education programmes, other resources may not (Pinter, 2005).

While archaeological sites and remains have long been an important foundation of heritage tourism, archaeological tourism specifically developed because of increased awareness and demand emanating from society towards the past and its remains, especially in the last three decades. Several studies have identified the aspects of archaeology that are of interest to the public, in the United States, Canada and Great Britain in particular (Pokytolo, 2002; Pokytolo & Guppy, 1999; Ramos & Duganne, 2000). The interest attributed to archaeology is its capacity to inform about the way people lived before and its ability to connect the past with the present and the future. In spite of these studies, tools to identify the public and qualify their perceptions about archaeological heritage are scarce. Most heritage consumers consider it necessary to protect sites and archaeological remains, and it is important to consider the diversity of archaeology consumers, including tourists (domestic and international), local inhabitants and other individuals (e.g. researchers, experts, educators and administrators).

Actually, this diverse view of the public lies at the heart of the problem in the relations between archaeology and tourism. Indeed, the apparent symbiosis between archaeology and tourism hides a more difficult relationship. In particular, archaeologists are not always in favour of tourism's use of archaeological resources, considering that tourists and tourism infrastructure are essentially a threat to archaeological sites, frequently causing physical damage, as well as a potential loss of cultural significance. Damage to irreplaceable material culture is not only direct, as when remains are disordered, altered, destroyed or looted, but often an indirect result of poorly planned tourism development and its infrastructure, such as hotels, restaurants, roads and shops. These can drastically and permanently alter the cultural and physical environments of places (Pinter, 2005).

This view also reflects an exclusive appropriation by archaeologists of places and objects that are not in fact exclusively archaeological. Archaeological sites can be simultaneously research laboratories, spaces of preservation, commemorative sites and places for leisure activities. For example, archaeological parks, including Petra in Jordan, Bibracte in France and Aubechies in Wallonia, Belgium, represent innovative forms of archaeological heritage development together with projects for regional economic development. These multiple uses can be distributed in specific spaces or stacked in the same space at the same time (Gillot, 2008).

The cohabitation of various practices and heritage uses is a potential source of tension, or even conflict, between various users. For instance, the site of Bosra in southern Syria is composed of inherited spaces that have been the centre of multiple appropriations. In the absence of a management plan, controlling the partition of these spaces is problematic and conflicts between the institutional actors, the archaeologists, tourism operators and local populations are frequent. If public authorities, international heritage institutions and archaeologists would work together to remedy the problems of abandonment, impoverishment, and the degradation of the traditional built environment, local inhabitants would better appreciate the ancient remnants beneath their feet and funding might become more available for excavations, monument restoration and tourism development (Gillot, 2011).

These conflicts are indicative of the tensions between the three main functions of a site: preservation and knowledge creation, economic development (tourism) and a leisure and social setting for residents. This discord may appear from symbolic appropriation, as in the case of Bosra. Tensions can result from the intrinsic 'dissonance' of heritage (Graham et al., 2000, 2005; Tunbridge & Ashworth, 1996) when more than one group claims the same heritage, when heritages are created and/or celebrated differently within the same groups, or when heritage becomes commodified for tourism.

Views of the Relationships between Archaeology and Tourism

The relationships between archaeology and tourism are far from being constant and universally approved or challenged. Three main approaches can be distinguished. First is a heterogeneous academic view, which for a long time was dominated by a criticism of the negative aspects of tourism, including the commercialization or commoditization of archaeological and cultural manifestations, which raises many concerns about authenticity within the nexus of archaeology and tourism (Holtorf, 2013). Traditionally, in the world of archaeology, the relationships between archaeology, heritage and economic development were downplayed, even antagonistic. However, with the development of more applied research and the movement toward 'archaeology for the public good', this austere stance began to soften to the point where it is even sometimes seen as a positive relationship (Shanks, 2004).

Second is a pragmatic managerial view, which is firmly in favour of utilizing archaeological resources for economic development but remains cognizant of tourism's positive and negative impacts, according to specific contexts. For decades, scholars have examined the positive and negative outcomes of tourism (Hall & Lew, 2009; Mathieson & Wall, 1982), and many have questioned the value of tourism, given its destructive potential in sociocultural, economic and environmental terms. These concerns paved the way for archaeology specialists also to question the relationships between archaeology and tourism (Comer & Willems, 2011). Through time, the concept of heritage, including archaeology, evolved from a strictly cultural, academic and institutional definition to a pluralistic and socioeconomic one. Because heritage tourism is regarded as a socioeconomic, political and cultural phenomenon, the participation of archaeologists and archaeological resources in the development of societies is a significant consideration.

Relatedly, the book *Marketing Heritage* (Rowan & Baram, 2004) focuses on the role national archaeology and archaeology-based tourism may play in the processes of nationalism, as well as unification and pacification in regions and nations in a state of conflict (Kohl, 2004). The book's contributors insist that the role played by archaeologists in the commodification ('new global marketing') of the past, and more precisely the contribution of their excavations and writings and analyses to the heritage and tourist industries is vital. They recognize nevertheless that the relations between archaeology, heritage and tourism are problematic. They evoke in particular the problem of the appropriation of the past and of the erroneous images that are given to tourists and cultural actors. They also highlight the ambiguous impacts of the cultural and tourist valuation of sites and archaeological objects, in terms of authenticity on the one hand, and economic profitability on the other (Kohl, 2004). Even if the book offers a relatively critical evaluation of the relations between archaeology and tourism, it results in practical recommendations that aim to reconcile, not divide, these two domains. The book's contributors indeed argue that their participation in displaying and interpreting their excavations and research guarantees their autonomy and the accuracy of the information presented to the public. Similarly, a special issue of the

SAA Archaeological Record (Pinter, 2005) on heritage tourism arose out of a firm belief that archaeologists must play an advocacy role in archaeological tourism.

The third main approach to viewing the relationships between tourism and archaeology is a particularly 'utopian' institutional view of the beneficial effects of the valuation of heritage. Many studies concerning the relationships between culture, tourism and development have been commissioned by international agencies such as UNESCO and the World Bank. For instance, ICOMOS' International Scientific Committee on Archaeological Heritage Management (ICAHM) has produced a series of publications that examine how growing visitation has affected the historical and scientific values of one of the most famous archaeological World Heritage Sites: Petra, Jordan (Comer, 2012). These studies provide empirical cases of successes and failures, but readers should be cautious about their conclusions and their value judgments.

Given these issues outlined above, as many archaeologists desire to become more involved in democratizing archaeological heritage, what is the position of archaeologists? How can their reluctances about, or interest in, heritage tourism be understood? To answer these questions, we should return to the basics of archaeology and consider the epistemological, ethical and technical evolutions that have marked the discipline in recent decades.

The Archaeological Debate: Confrontation or Collaboration?

In the 19th century, the discipline of archaeology faced the instrumentalization and reinterpretation of its research approaches. This revamping of the scientific discourse took place at first within the framework of the identities and territorial claims of nation states, followed by the emergence of new states that arose from the process of decolonization. Archaeology has therefore been instrumental in supporting imperialism, nationalism and anticolonialism between the 19th century and the second half of the 20th century. From the 1960s, archaeology was called to intervene in the construction of heritage, its protection and its valuation, at the same time as the adoption of policies and standards of heritage protection at the international and national level. The development of rescue archaeology and legislation concerning heritage, land use and urban planning, both in developed and developing countries, also spurred archaeologists to work with public authorities and private actors. Simultaneously, the growth and democratization of leisure activities, including tourism, after World War II, aroused an increased interest within society for history, archaeology and heritage (Schnapp, 1993, 2002; Silberman, 1989; Trigger, 1995).

Despite these changes, archaeologists seemed to remain disinterested in their relationships with the public and society in a broad sense. This reluctance probably demonstrated the fundamental character of archaeological science and the difficulty it faced in providing practical knowledge that the public could consume. It also reflects the seemingly insurmountable paradox between the principle of protecting archaeological remains and that of providing access to them for large numbers of people. Access would lead to high tourist visitation at certain archaeological sites and museums and would thus become a source of pressure on the remains and a constraint to the work of archaeologists. Indeed, the very presence of tourists, particularly in large groups, is regarded as an inconvenience by archaeologists, either because visitors might damage or destroy important findings and information, or because they interrupt their work by asking questions. Many archaeologists may therefore feel that they themselves are 'tourist attractions' rather than hard-working scientists. Archaeologists' reluctance to communicate with visitors stems largely from the fact that time on excavations is limited and therefore precious, especially when their projects are of a short-term nature. On the contrary, the visiting public hopes to understand the history being uncovered and witness the progress being made, leading to disappointment by a lack of explanation. These conflicting views are most prominent in certain sites that suffer from excessive numbers of visitors: Angkor in Cambodia, Lascaux in France, the Valley of the Kings in Egypt, Machu Picchu in Peru and Pompeii in Italy are frequently noted examples.

However, the reluctance, or even the refusal, on the part of archaeologists to consider the current implications of their labour and to participate in the development of heritage tourism seem inappropriate if we consider the evolution of heritage management since the 1990s. This evolution was characterized by a move from a prevailing philosophy of strict protection (involving closures or strict limitations on access to archaeological sites) to a more dynamic approach that advocates accessibility to as many people as possible. The most recent charters affecting archaeological resource management emphasize the importance of developing a system that protects the archaeological heritage, integrates its values, and aims for dialogue between stakeholders (e.g. archaeologists, planners, public authorities, resident populations and tourists).[1]

Attempts at reconciliation between research, conservation and visitation have particularly complex implications for the discipline of archaeology. Indeed, archaeologists are not only called to participate actively in developing their research but also in valuating the archaeological heritage (McManamon, 1991). Furthermore, the evolution of archaeology in the 1970s and 1980s favoured greater reflexivity, brought about by the will of archaeologists to play a greater social role and participate in the processes of heritagization and 'touristification' of archaeological sites and objects. This change, albeit remaining within the scope of rescue and preventive archaeology, has influenced the academic discourse to become less punitive towards the impacts of tourism development at archaeological sites (Djindjian, 2010). These changes are part of broader changes in communicating the archaeological message and in the image of the discipline, both between archaeologists (need for communication between researchers) and between archaeologists and the public (need to inform the public and legitimize the funds needed for research) (Schnapp, 2002). As a scientific discipline, liberal profession, or an administrative function focused on resource management, archaeology has become an activity embraced by many different stakeholders (Firth, 1995).

This evolution has been particularly evident in the Anglo-Saxon world, the Scandinavian countries and more recently, in the Spanish-speaking and French-speaking world. Indeed, since the 1970s, the emergence of preventive archaeology and the professionalization of the discipline have stimulated the development of initiatives in favour of protecting archaeological resources and public participation. For example, the French National Institute for Preventive Archaeological Research (INRAP), or research centres such as the CreA-Patrimoine in Belgium,[2] have actively developed conferences, lessons and multimedia tools to explain archaeological work to larger public audiences. Since the 1990s, the post-processual movement also emphasized the importance of the social and public dimension of archaeology, either considered as an objective science or as a discourse and a mode of cultural production (Hodder, 1995; Hodder *et al.*, 1995; Russell, 2006[3]; Shanks, 2004).

Public archaeology, community-sponsored archaeology and archaeological heritage management are examples of many sub-disciplines that have emerged out of an interest in protecting and interpreting heritage on the one hand, and involving the public and local communities in these processes, on the other hand (Hodder, 2000). First of all, the concept of 'public archaeology' was introduced by Charles McGimsey in 1972 to demonstrate the way the discipline should engage with social, political and economic questions (McGimsey, 2004). Public archaeology is based on the idea that archaeology is, by nature, a public activity or public good, a source of knowledge, as well as an object of manipulation and erroneous representations. It thus deserves, as such, to be supported by the state. The objective of public archaeology is in particular to dispel the image of the intrepid adventurer looking for hidden treasures, replacing it with an image that archaeologists wish to portray. In addition, public archaeology aims to involve the public in constructing the past as a means of appeasing the public's interest and increasing its awareness of the need to protect the archaeological heritage (Darvill, 2004; Little, 2002; Merriman, 2004; Schadla-Hall, 1999; White *et al.*, 2004). Finally, the idea of public archaeology was formulated in the United States and Australia within the framework relationship building between archaeologists and native communities (Berggren & Hodder, 2003; Cipolla *et al.*, 2019; McNiven, 2016). The notions of collaborative research and community-sponsored archaeology refer to participatory steps in involving communities in interpreting and developing their pasts (Layton, 1994; Shackel & Chambers, 2004).

Another sub-discipline, archaeological resource management (ARM), emerged in the United States in the second half of the 1970s to answer three objectives: maintain the diversity of archaeological remains existing on an area, make the archaeological heritage accessible to cultural consumers and limit the conflicts between varying uses of the land containing archaeological remains (Berggren & Hodder, 2003; Cleere, 1989; Cooper *et al.*, 1995; Mathers *et al.*, 2005; McManamon & Hatton, 2000). ARM is a field of archaeology that is defined as 'the protection and the administration of the archaeological heritage in its original environment and in its relation in the history and in the contemporary society' (Carman, 2000). It includes activities such as inventorying, excavating and researching, as well as protecting, presenting and educating (Biörnstad, 1989; Mayer-Oakes, 1989). Nevertheless, the recognition of ARM as a sub-discipline of archaeology is uneven in different parts of the world (Carman, 2000; Kristiansen, 1989; Tainter, 2004; Willems *et al.*, 2018).

French-speaking Europe, in particular France, was slower to deal with the questions of developing heritage and the social role of archaeology (Demoule & Stiegler, 2008). Since the 1980s, the debate centred around two issues. First was that of the erroneous image of French archaeology within political circles, with planners and in the public eye. In the French context, archaeology was dominated by the image of adventurers in search of rare treasures (Charpentier, 2002; Demoule, 2002, 2005; Pesez, 1997). Secondly, the unease of the profession was magnified by the dichotomy between the prosaic image of metropolitan archaeology (practised in continental France) and the public image of exoticism associated with extrapolitan archaeology (the big excavations abroad).

From these debates, the idea of opening archaeology to the public appeared as a way of mitigating the risk of confining the discipline from the society that supports it (Demoule, 2007). As part of this change and to fulfil their growing social obligations, archaeologists began to engage with town and country planning (Demoule & Stiegler, 2008). This led to an examination of the role of archaeologists in national and regional development, including physical infrastructure planning and development, but also tourism development, as most countries and regions now see tourism as a key socioeconomic development tool. Tourism, however, continued to be viewed with suspicion because of its potential for the 'Disneyfication' of archaeological sites and heritage places.

The cross-cutting issue in all of these concerns is about the independence of archaeology vis-à-vis political and economic interests. Pointing to this is the division between traditional archaeology and applied archaeology, which seems to continue to grow. The former supports a neutral and detached science, whereas the latter calls on archaeologists and historians to participate actively in the valorisation of their research as a public, common good.

Regardless of the evolving position towards tourism, most archaeologists remain reluctant to assume the tasks of visitor management and tourism development for a variety of reasons. First, these tasks must deal with the problem of interpreting and popularizing a knowledge base that is constantly being revised and updated. Secondly, these tasks require marketing efforts that identify diverse public consumer segments and cater to their needs. Third, traditional archaeological programs and archaeology as a discipline remains largely unfamiliar with the principles of management (Darvill, 1995). Fourth, archaeologists are keen to preserve their independence from political and the other forms of pressure. The fifth source of reluctance is that collaborating with the other partners – planners and tourism actors in particular – requires a common 'language', which neither side is yet equipped to speak. Sixth, financing visitor operations is an obstacle that most do not desire to deal with; it is hard enough to seek funding for their hands-on archaeological activities. Finally, relatively few academic archaeologists have adapted to the contemporary changes in the discipline. Indeed, most archaeologists today lead studies of art history and archaeology where preservation and asset management are rarely taught (Stanley Price, 1989).

The archaeological community recognizes archaeologists' role as encompassing three functions (Demoule, 2002). The first one is a conserver. Archaeologists must protect the remains of the past and arbitrate the contradictory interests of urban and regional planning. The second role is historic; the archaeologist must reconstitute and restore the past. The third is cultural. The archaeologist is tasked with assuring the transmission and distribution of knowledge. To sum up the statement of Mayer-Oakes (1989), the archaeologist is called to play a triple role: scholar, steward and storyteller. Since a relationship between archaeology and tourism seems possible, it is therefore advisable at this stage to consider the modalities of this collaboration.

Bridging the Gap, Engaging the Public: Archaeologists Working for and with Tourists

In spite of the differences between Francophone and Anglophone traditions, and applied and traditional archaeology, the academic and professional worlds seem to have a common view with respect to the responsibilities of archaeologists towards archaeological sites and data. In this context, the development of an archaeological site for tourism may be seen from two dimensions. The first one relates to the conservation and presentation of the research, while the second one deals with the restoration and the destination of archaeological remains. Both involve measures during and after the research, as well as collaboration with other experts (Pesez, 1997). Thus, the protection of excavated remains aims to protect the information for future research (Berducou, 1980). The preservation is justified by the necessity of answering current or future problems, as well as by the 'perishable' nature of the archaeological remains. Finally, it is recognized that the duty of scientists is to explain the nature and the impact of their work, both to their colleagues and to the public. Most professional archaeologists see opportunities to collaborate between tourism and archaeological heritage (McManamon, 1991; Pinter, 2005). They believe that a balance can be reached between conservation and tourism by multiplying communication efforts between interest groups.

To make archaeology accessible, three main modalities are considered: opening sites and museums for increased visitation, public access to current excavations, and public participation in excavations and laboratory work. Virtual techniques also play an important role in this mix. Most heritage site administrators see their role as managing for the broader public, not just for tourists. Public participation can be enhanced by opening archaeological digs to nonqualified volunteers. The renewal of the relationship between archaeological work and tourism stems from a broad concern for the visitor experience, and the potential impact of visitors on the archaeological record. The adequacy of resource conservation, sustainability and management; the appropriateness of public access and associated site improvements; the interpretation of archaeological resources; and the economic viability of open sites and visitor facilities remain centres of debate among archaeologists and heritage and tourism professionals. The development of information technology-based tools for interpreting and presenting the results of archaeological studies can foster a broader understanding and support heritage tourism initiatives (Hausmann & Weuster, 2018; Solima & Izzo, 2018).

The Anglo-Saxon tradition of social archaeology places interpretation at the centre of this approach (Shanks, 2004). In the tourism literature, the notion is understood more as shaping or reproducing the past, rather than purely describing it (Moscardo, 1996; Nuryanti, 1996; Timothy, 2011). The effectiveness of interpretation is determined by visitors' participation and interaction, as well as by the pedagogical approaches and interpretive methods. Interpretation is thus a form of mediation that aims to establish emotional and intellectual contact between the public and the interpreted objects, but also to put objects into context (Moscardo, 1996). For example, the archaeosite of Aubechies, Belgium, provides animation for youth and adult visitors – reconstructions of prehistoric and protohistoric houses with demonstrations by crafters, and a shop selling some of their crafts (e.g. Celtic jewellery). The archaeological park of Bibracte, France, also proposes thematic walks and visits, workshops and handicrafts, while remaining an active excavation site.

To summarize, Melotti (2011) considers that engaging visitors actively in archaeology under controlled and limited conditions may also be appropriate. Such 'participatory archaeology' can offer opportunities for educating the public about archaeology and resource stewardship, as well as providing a rewarding leisure activity. Finally, archaeotourism could also have significant implications for indigenous people and other groups.

Archaeology-based tourism has the potential of not only bringing financial gains but also helping to create a more cohesive identity within local or descendant communities. In this context, the role of the archaeologist in developing sustainable and responsible archaeological tourism is crucial. Pacifico and Vogel (2012) suggest that when tourism and archaeological sites meet, archaeologists should avoid arguments about ownership and act as facilitators of an open dialogue about the consequences of tourism development for all parties involved.

Conclusion

Since the 1970s, archaeology has undergone important changes that have required the discipline to deal with public questions and contribute to the public good, particularly concerning the conservation and development of archaeological sites and heritage. The professionalization and the privatization of archaeology have also led to new forms of interaction between archaeologists and tourists. The needed collaboration among archaeologists, politicians, planners, tourism entrepreneurs, visitors and local communities may still be far from being realized. Nonetheless, this chapter shows that the epistemological, ethical and technical developments within the discipline of archaeology, along with new perspectives for more sustainable forms of heritage tourism, respectful of archaeological resources, lay the foundations for a new space of dialogue. The relationship between archaeology and tourism is no longer a polarized matter of confrontation or symbiosis but a matter of collaboration between two activities sharing the same spaces and resources and assuming the transmission of knowledge to future generations. To conclude, this chapter underscores the benefits of the involvement of archaeologists in archaeological resource management and maintaining a public presence, as well as the intervention of private players and civil society in the definition, management and development of archaeological heritage-based tourism. In this respect, it seems prudent to foster communication between stakeholders and to make tourists and local communities aware of the multiplicity of values of archaeology and its activities.

Notes

(1) For example, the 'Recommendation concerning the protection and enhancement of archaeological heritage in the context of town and country planning operations' (Council of Europe, 1989) is based on the idea that 'the protection and the development of the archaeological heritage constitutes a factor of at the same time cultural, tourist and economic development'. The Charter for the Protection and Management of the Archaeological Heritage (ICOMOS, 1990) and the European Convention on the Protection of the Archaeological Heritage (Council of Europe, 1992) also underline that the protection of heritage implies its identification, protection, study, restoration and development. Development may be defined as increasing public access and

exhibitions. Finally, the ICOMOS Charter for the Interpretation and the Presentation of Cultural Heritage Sites in 2007 provides a systematization of the techniques and standards of presentation and interpretation, as well as a definition of professional and ethical directives.
(2) The CReA-Patrimoine is a leading research centre at the Université libre de Bruxelles that promotes national and international programs on archaeology and cultural heritage. It constitutes the privileged partnership between the university and the public authorities in charge of cultural heritage.
(3) These authors consider that archaeology has the power to influence the production and distribution of images of the past and that archaeologists have the duty to contribute to more accurate knowledge and presentation of the latter. It is also a condition of their autonomy.

References

Baram, U. (2008) Tourism and archaeology. In D.M. Pearsall (ed.) *Encyclopedia of Archaeology* (pp. 2131–2134). Amsterdam: Elsevier.
Biörnstad, M. (1989) The ICOMOS International Committee on Archaeological Heritage Management (ICAHM). In H. Cleere (ed.) *Archaeological Heritage Management in the Modern World* (pp. 70–78). London: One World Archaeology.
Berggren, A. and Hodder, I. (2003) Social practice, method, and some problems of field archaeology. *American Antiquity* 68 (3), 428–429.
Berducou, M. (1980) La conservation archéologique. In A. Schnapp (ed.) *L'archéologie aujourd'hui* (pp. 149–169). Paris: Hachette Littérature.
Carman, J. (2000) Theorising the practice of archaeological resource management. *Archaeologia Polona* 38, 5–21.
Charpentier, V. (2002) Image réelle, image rêvée: Quel visage pour l'archéologie? *Raison présente* 142, 9–14.
Cipolla, C.N., Quinn, J. and Levy, J. (2019) Theory in collaborative indigenous archaeology: Insights from Mohegan. *American Antiquity* 84 (1), 127–142.
Cleere, H. (ed.) (1989) *Archaeological Heritage Management in the Modern World*. London: One World Archaeology.
Comer, D.C. (2012) *Tourism and Archaeological Heritage Management at Petra: Driver to Development or Destruction?* New York: Springer.
Comer, D.C. and Willems, W.J.H. (2011) Tourism and archaeological heritage: Driver to development or destruction? In *Proceedings of the 17th ICOMOS General Assembly Symposium Organised by ICOMOS France November 27 – December 2, 2011, Paris*, pp. 506–518.
Cooper, M.A., Firth, A., Carman, J. and Wheatley, D. (eds) (1995) *Managing Archaeology*. London: Routledge.
Council of Europe (1989) Reference texts. Online at: https://www.coe.int/en/web/culture-and-heritage/texts-of-reference (last accessed February 27, 2020).
Council of Europe (1992) Details of Treaty No. 143. Online at: https://www.coe.int/en/web/conventions/full-list/-/conventions/treaty/143 (last accessed February 27, 2020).
Darvill, T. (1995) Preparing archaeologists for management. In M.A. Cooper, A. Firth, J. Carman and D. Wheatley (eds) *Managing Archaeology* (pp. 175–187). London: Routledge.
Darvill, T. (2004) Public archaeology: A European perspective. In J.L. Bintliff (ed.) *A Companion to Archaeology* (pp. 409–433). Oxford: Blackwell.
Demoule, J.-P. (2002) L'archéologie dans la société: les responsabilités des archéologues. In J.-P. Demoule (ed.) *Guide des méthodes de l'archéologie* (pp. 232–248). Paris: La Découverte.
Demoule, J.-P. (2005) *L'archéologie, entre science et passion*. Paris: Gallimard.

Demoule, J.-P. (ed.) (2007) *L'archéologie préventive dans le monde: Apports de l'archéologie préventive à la connaissance du passé*. Paris: La Découverte.

Demoule, J.-P. and Stiegler, B. (2008) *L'avenir du passé: Modernité de l'archéologie*. Paris: La Découverte.

Djindjian, F. (2010) Le rôle de l'archéologue dans la société contemporaine. *Diogène* 2010 (229–230), 78–90.

Firth, A. (1995) Ghosts in the machine. In M.A. Cooper, A. Firth, J. Carman and D. Wheatley (eds) *Managing Archaeology* (pp. 51–67). London: Routledge.

Gillot, L. (2008) La mise en valeur des sites archéologiques: un rapprochement entre archéologie, tourisme et développement: Le cas de la Syrie (Developing Archaeological Sites: Bridging the gap between archaeology, tourism and development: The Syrian case). Unpublished Thesis, Brussels.

Gillot, L. (2011) Les dynamiques socio-spatiales de mise en valeur des sites archéologiques habités: le cas de Bosra en Syrie. In I. Backouche, F. Ripoll, S. Tissot and V. Veschambre (eds) *La dimension spatiale des inégalités: Regards croisés des sciences sociales* (pp. 141–161). Rennes: PUR.

Giraudo, R.F. and Porter, B.W. (2010) Archaeotourism and the crux of development. *Anthropology News* 51 (8), 7–8.

Graham, B., Ashworth, G.J. and Tunbridge, J.E. (2000) *A Geography of Heritage: Power, Culture and Economy*. London: Arnold.

Graham, B., Ashworth, G.J. and Tunbridge, J.E. (2005) The uses and abuses of heritage. In G. Corsane (ed.) *Heritage, Museums and Galleries: An Introductory Reader* (pp. 26–37). London: Routledge.

Hall, C.M. and Lew, A.A. (2009) *Understanding and Managing Tourism Impacts: An Integrated Approach*. London: Routledge.

Hausmann, A. and Weuster, L. (2018) Possible marketing tools for heritage tourism: The potential of implementing information and communication technology. *Journal of Heritage Tourism* 13 (3), 273–284.

Hodder, I. (1995) *Theory and Practice in Archaeology*. London: Routledge.

Hodder, I. (ed.) (2000) *Towards Reflexive Method in Archaeology: The Example at Çatalhöyük*. Cambridge: McDonald Institute for Archaeological Research.

Hodder, I., Shanks, M., Alexandri, A., Buchli, V., Carman, J., Last, J. and Lucas, G. (1995) *Interpreting Archaeology: Finding Meaning in the Past*. London: Routledge.

Holtorf, C. (2013) On pastness: A reconsideration of materiality in archaeological object authenticity. *Anthropological Quarterly* 86 (2), 427–443.

ICOMOS (1990) Charter for the Protection and Management of the Archaeological Heritage. Online at: http://icahm.icomos.org/wp-content/uploads/2017/01/1990-Lausanne-Charter-for-Protection-and-Management-of-Archaeological-Heritage.pdf (last accessed February 27, 2020).

Kohl, P.L. (2004) Making the past profitable in an age of globalization and national ownership: Contradictions and considerations. In Y. Rowan and U. Baram (eds) *Marketing Heritage: Archaeology and the Consumption of the Past* (pp. 295–301). Walnut Creek, CA: AltaMira.

Kristiansen, K. (1989) Perspectives on the archaeological heritage: History and future. In H. Cleere (ed.) *Archaeological Heritage Management in the Modern World* (pp. 23–29). London: Unwin Hyman.

Layton, R. (ed.) (1994) *Who Needs the Past? Indigenous Values and Archaeology*. London: Routledge.

Little, B.J. (2002) *Public Benefits of Archaeology*. Gainesville: University Press of Florida.

Mathers, C., Darvill, T. and Little, B.J. (2005) *Heritage of Value, Archaeology of Renown: Reshaping Archaeological Assessment and Significance*. Gainesville: University Press of Florida.

Mathieson, A. and Wall, G. (1982) *Tourism: Economic, Physical and Social Impacts*. London: Longman.

Mayer-Oakes, W.J. (1989) Science, service and stewardship: A basis for the ideal archaeology of the future. In H. Cleere (ed.) *Archaeological Heritage Management in the Modern World* (pp. 52–58). London: One World Archaeology.
McManamon, F.P. and Hatton, A. (2000) *Cultural Resource Management in Contemporary Society: Perspectives on Managing and Presenting the Past*. London: Routledge.
McManamon, F. (1991) The many publics for archaeology. *American Antiquity* 56 (1), 123–127.
McNiven, I.J. (2016) Theoretical challenges of indigenous archaeology: Setting an agenda. *American Antiquity* 81 (1), 27–41.
McGimsey, C.R. (2004) *CRM on CRM: One Person's Perspective on the Birth and Early Development of Cultural Resource Management*. Fayetteville: Arkansas Archaeological Survey.
Merriman, M. (2004) *Public Archaeology*. London: Routledge.
Melotti, M. (2011) *The Plastic Venuses: Archaeological Tourism in Post-Modern Society*. Newcastle: Cambridge Scholars Publishing.
Moscardo, G. (1996) Mindful visitors: Heritage and tourism. *Annals of Tourism Research* 23 (2), 376–397.
Nuryanti, W. (1996) Heritage and postmodern tourism. *Annals of Tourism Research* 23 (2), 249–260.
Pacifico, D. and Vogel, M. (2012) Archaeological sites, modern communities, and tourism. *Annals of Tourism Research* 39 (3), 1588–1611.
Pesez, J.-M. (1997) *L'archéologie: Mutations, Missions, Methods*. Paris: Nathan.
Pinter, T.L. (2005) Heritage tourism and archaeology: Critical issues. *SAA Archaeological Record* 5 (3), 9–11.
Pokytolo, D. (2002) Public opinion and Canadian archaeological heritage: A national perspective. *Canadian Journal of Archaeology* 26 (2), 88–129.
Pokytolo, D. and Guppy, N. (1999) Public opinion and archaeological heritage: Views from outside the profession. *American Antiquity* 64 (3), 400–416.
Ramos, M. and Duganne, D. (2000) *Exploring Public Perceptions and Attitudes about Archaeology*. Rochester, NY: Harris Interactive.
Rowan, Y. and Baram, U. (eds) (2004) *Marketing Heritage: Archaeology and the Consumption of the Past*. Walnut Creek: AltaMira.
Russell, I. (ed.) (2006) *Images, Representations and Heritage: Moving beyond Modern Approaches to Archaeology*. New York: Springer.
Shackel, P.A. and Chambers, E.J. (2004) *Places in Mind: Public Archaeology as Applied Anthropology*. London: Routledge.
Schadla-Hall, T. (1999) Editorial: Public archaeology. *European Journal of Archaeology* 2 (2), 147–158.
Schnapp, A. (1993) *La conquête du passé: Aux origines de l'archéologie*. Paris: Editions Carré.
Schnapp, A. (2002) Histoire de l'archéologie et l'archéologie dans l'histoire. In J.-P. Demoule (ed.) *Guide des méthodes de l'archéologie* (pp. 9–38). Paris: La Découverte.
Shanks, M. (2004) Three rooms: Archaeology and performance. *Journal of Social Archaeology* 4, 147–179.
Silberman, N. (1989) *Between Past and Present: Archaeology, Ideology and Nationalism in the Modern Middle East*. New York: Anchor Books.
Solima, L. and Izzo, F. (2018) QR codes in cultural heritage tourism: New communications technologies and future prospects in Naples and Warsaw. *Journal of Heritage Tourism* 13 (2), 115–127.
Stanley Price, N.P. (1989) Archaeology and conservation training at the international level. In H. Cleere (ed.) *Archaeological Heritage Management in the Modern World* (pp. 292–301). London: One World Archaeology.

Tainter, J.A. (2004) Persistent dilemmas in American cultural resource management. In J.L. Bintliff (ed.) *A Companion to Archaeology* (pp. 435–453). Oxford: Blackwell.

Timothy, D.J. (2011) *Cultural Heritage and Tourism: An Introduction*. Bristol: Channel View Publications.

Trigger, B.G. (1995) Romanticism, nationalism and archaeology. In P.L. Kohl and C. Fawcett (eds) *Nationalism, Politics and the Practice of Archaeology* (pp. 263–279). Cambridge: Cambridge University Press.

Tunbridge, J.E. and Ashworth, G.J. (1996) *Dissonant Heritage: The Management of the Past as a Resource in Conflict*. Chichester: Wiley.

Walker, C. and Carr, N. (eds) (2013) *Tourism and Archaeology: Sustainable Meeting Grounds*. Walnut Creek: Left Coast Press.

White, N.M., Weisman, B.R., Tykot, R.H., Wells, E.C., Davis-Salazar, K.L., Arthur, J.W. and Weedman, K. (2004) Academic archaeology is public archaeology. *The SAA Archaeological Record* 4 (2), 26–29.

Willems, A., Thomas, A., Mena, A.C., Čeginskas, V., Immonen, V., Kalakoski, I., Lähdesmäki, T., Lähdesmäki, U., Gowen-Larsen, M., Marciniak, A., Pérez Gonzáles, E., White, C. and Mazel, A.D. (2018) Teaching archaeological heritage management: Towards a change in paradigms. *Conservation and Management of Archaeological Sites* 20 (5–6), 297–318.

Wurz, S. and van der Merwe, J.H. (2005) Gauging site sensitivity for sustainable archaeo-tourism in the Western Cape Province of South Africa. *The South African Archaeological Bulletin* 60 (181), 10–19.

3 Tourism and the Economic Value of Archaeology

Paul Burtenshaw

Introduction

Archaeological sites and materials possess economic value or economic capital – the ability to generate financial benefits through a variety of activities (Burtenshaw, 2014; Mason, 1999). Despite its widespread use, this ability has an uncomfortable relationship with archaeology's other capitals or values. Economic value is often seen as distinct from, and indeed threatening to, cultural heritage's social and cultural values. This tension is perhaps most intense when considering its use as a tourism resource, which some may view as a necessary burden to be able to support the preservation of archaeology and attract funds and attention from stakeholders to support other activities. While the tensions between different uses of archaeology are understandable, this chapter suggests that a binary view is unhelpful and unnecessary. The position of archaeology's economic capital is reviewed, its relationship with other values discussed, and empirical material from Kilmartin Glen in Scotland is presented to demonstrate the complex and inter-dependent relationship of archaeology's values. The economic uses of archaeology, including through tourism, should be seen in a complex relationship with its other values which can be positive or negative depending on management.

The Economic Value of Archaeology

A major part of cultural heritage management is understanding why people may find a particular tangible or intangible item of heritage important. Values have been the main analytical tool to assess the significance and qualities of heritage, attempting to break down the overall significance or importance of heritage into different characteristics (Mason & Avarami, 2002). Exactly how to categorize the significance of heritage, including archaeological remains, and divide its various characteristics into values has been a matter of debate with several typologies suggested

(Mason, 2008; Throsby, 2001; Timothy & Boyd, 2003). A selection of values that have been proposed include aesthetic value, informational or scientific value, symbolic or identity value, spiritual value and historical value to name a few. These values can overlap and have different meanings depending on the typology used.

One suggested reason people might find heritage valuable or important is that it generates economic benefits (e.g. jobs or regional revenue) for them. This is most often called 'economic value'. This term is problematic, and I have argued elsewhere for it to be called 'economic capital' (Burtenshaw, 2014), but the former term will be retained for this chapter. Cultural heritage can act as a fiscal resource by stimulating economic impacts through various activities related to it (Çela et al., 2009; Nijkamp, 2012; VanBlarcom & Kayahan, 2011). Destinations (countries, cities, regions) throughout the world target cultural heritage for its economic value. Lipe (2009: 61), in his value scheme, notes that such a value can be positive or negative depending on if the site can generate monetary benefits or might be 'in the way' of other economic uses of a place.

The fact that individuals or communities might find the economic value of cultural heritage important is widely acknowledged. It is regularly stated that archaeology's economic value is essential to its public support and a significant reason why it is important to various stakeholders (e.g. Flatman, 2012; Hodder, 2010; Selvakumar, 2010). In fact, its economic value often underscores the justification of heritage preservation. This includes at the national level where negotiations for budgets and government attention are often conducted with economic-based evidence (Belfiore, 2012; Bewley & Maer, 2014), at the international level with heritage organizations keen to trumpet heritage's potential contribution to sustainable development (both economic and non-economic) (ICOMOS, 2017; UNESCO, 2010), and at the local level with archaeologists situating local relationships with archaeology around job creation and economic benefits (Burtenshaw & Palmer, 2014; Gould, 2017; Labadi & Gould, 2015). Despite, or perhaps because of, this acknowledged importance of archaeology's ability to generate economic benefits, it is often separated from, or put into opposition to, its other values.

While Lipe (2009) includes pecuniary value in his scheme of various values, perhaps the most influential statement on values, the Burra Charter (Australian ICOMOS, 2013; Lafrenz Samuels, 2008), purposefully excludes economic value as a reason for preservation. Efforts to conceptualize how to communicate the value of heritage in the UK have proposed the separation of 'intrinsic' and 'instrumental' benefits (Holden, 2006), where social and economic effects are considered ancillary to cultural benefits. Carver's (1996) value scheme pits archaeology's informational worth against the forces of markets and politics. While public archaeologists have stated the economic importance to public value, there is comparatively little attention paid to the theme in overviews of the subject

(Burtenshaw, 2014). Of course, different value schemes have very different agendas to meet and tasks to perform, but this admittedly cursory summary highlights a background of discomfort about whether people valuing heritage in economic terms is a legitimate or worthy reason. As Graham *et al.* (2000: 129) encapsulate:

> Historically, the economic functions of heritage have generally been presented as subsequent or secondary and often barely tolerated uses of monuments, sites and places, which have been initially identified, preserved and interpreted for quite other reasons.

The reasons for this discomfort are manifold. As Mason and Avrami (2002) note, all the values of heritage are constantly changing and often in conflict. However, this contention seems heightened where economic value is concerned. There is a common idea among cultural specialists that economics and culture are incompatible philosophies (Mason, 1999). Economists are regarded as attempting to reduce all decisions into market value, while those in opposition, including cultural purists, have been accused of not living in the 'real world'. Attempts have been made to close this conceptual divide by both economists and archaeologists (e.g. Carman, 2002; Darvill, 1995; Mourato & Mazzanti, 2002), but their ideas and methods have largely failed to gain traction, even though the close relationship between these approaches is sometimes acknowledged (Mason, 2002; Throsby, 2002). Part of those efforts has been the development of measurement techniques that attempt to quantify all heritage values, as the more understandable and communicable economic data have been seen to overpower the data advocating for other values. As a result, many authors have identified a traditional distaste for economics within archaeology and related fields (e.g. Carman, 2005; Lafrenz Samuels, 2009).

More practically, and perhaps at the forefront in the minds of archaeologists and heritage managers, is the fear that the desire for tangible economic outcomes will trump the development or preservation of other sociocultural benefits (Graham *et al.*, 2000; Lipe, 1984; Mason, 1999). Perhaps the most obvious case in point is in the antiquities trade. Here, the monetary worth is pitched as the most dangerous value, because it is short sighted and frequently results in looting and other forms of irreparable damage. It is considered more desirable for people to appreciate archaeology for other, less-destructive reasons.

Such fears become clear in discussions about tourism and archaeology. Archaeology's economic value can manifest in many ways. Artefacts are sold or traded on the arts market (Brodie, 2014), although even 'legal' sales often raise ethical questions (Massy, 2008). Heritage can be re-used, such as in the repurposing of historic buildings for contemporary uses (Assefa & Ambler, 2017; Johnson & Thomas, 1995; Lipe, 1984; Timothy, 2011). Conservation activities create jobs and secondary economic benefits (Klamer & Zuidhof, 1999), and heritage is frequently used as part of branding or as a space for other events (Starr, 2010). However, tourism is by far the largest economic use of heritage (Throsby, 2002; Timothy, 2011).

Tourism generates an economic value through entrance fees to archaeological sites and museums, as well as the services visitors use in travelling to the destination, such as food, accommodation and transportation, as well the associated public spending on infrastructure development (Bowitz & Ibenholt, 2009). Tourism is the primary economic justification for archaeology throughout the world and makes up the lion's share of economic appraisals of archaeology (Bewley & Maeer, 2014; Castañeda & Mathews, 2013). The Global Heritage Fund (2010) implores people to help 'Save our Vanishing Heritage' by visiting. The organization calculates that heritage site-based tourism in the developing world alone could be worth $100 billion annually by 2025. Hall and McArthur (1996: 6) note that 'one of the main justifications for preserving heritage, especially from government and the private sector, is the value of heritage for tourism and recreation', a sentiment commonly expressed by heritage specialists (e.g. Adams, 2010; Gould, 2018; McManus, 1997).

Of course, the risks of tourism are well understood. Presenting an archaeological site as an attraction requires emphasising the characteristics of the locality that appeal to the tourist market (Silverman, 2002; Timothy & Boyd, 2003). The elements of archaeology that appeal to tourists may be different to those valued by local communities or stakeholders who already use the site for other purposes. The need to present an archaeological site so that it suits a particular market may require physical changes, such as restricted access or political or symbolic changes, such as certain aspects of the site's history being promoted over others (Merhav & Killebrew, 1998). The physical damage of excessive tourism monuments themselves is well known, with a variety of environmental impacts being brought to bear on ancient materials (Drdácký & Drdácký, 2006). Tourists' physical impacts are not limited to the relics and physical materials alone, for the infrastructural needs of tourism can also have major impacts on surrounding archaeological zones (Comer, 2012). And of course, heritage tourism comes with all of the wider environmental and cultural concerns of every other form of tourism (Scheyvens, 2002).

Seen within the wider context of the clash of economics and culture, and the problems that tourism effects, tourism is often seen by archaeologists and heritage managers as a Faustian bargain necessary to fund and support the 'true' aims of archaeology: discovery and research, preservation and education (Timothy, 2014). Some archaeologists may see broader benefits to local communities as an outcome of this bargain, perhaps as part of strategies to incentivize preservation, or supporting the ethical goals of local development (Burtenshaw & Palmer, 2014). However, several studies have pointed to the fact that local residents often do not benefit directly from tourism (Adams, 2010) and have relatively little voice in how tourism is developed (Aas *et al.*, 2005; Timothy, 2015).

However, the reality is of course more complex. While tourism can be a destructive force, its benefits beyond increased revenues are well documented, including intercultural dialogue, community wellbeing, and heritage and natural environmental preservation. Properly managed, revenue from tourism can reach relevant communities and help protect archaeological sites (Coben, 2014). However, it is important that the economic benefits of tourism do not overshadow the social and cultural benefits. The relationship between these values is complex and inter-related. The following section examines these ideas in relation to tourism at the archaeological remains in Kilmartin Glen as evidence of the way in which the economic and sociocultural values of archaeology can interact in positive, albeit complex, ways.

The Economic Value of Kilmartin Glen and Kilmartin House Museum

Kilmartin Glen (the Glen) is located within the region of mid-Argyll in western Scotland, approximately 100 miles northwest of Glasgow. The Glen (or valley) contains over 50 scheduled archaeological monuments, and surveys have revealed hundreds of prehistoric sites in the local area (Historic Scotland, 2007; RCAHMS, 1999). The area's archaeological remains include Neolithic (4000–2000 BC) chambered cairns, as well as a Neolithic and Bronze Age (2500–600BC) linear cemetery (Figure 3.1). Also associated with this period are cup-and-ring carvings on rock; this area includes the greatest concentration and largest sheets of these carvings in Europe. The landscape of the period also consists of various standing stones, stone circles and henges (Figure 3.2). Due to the density and

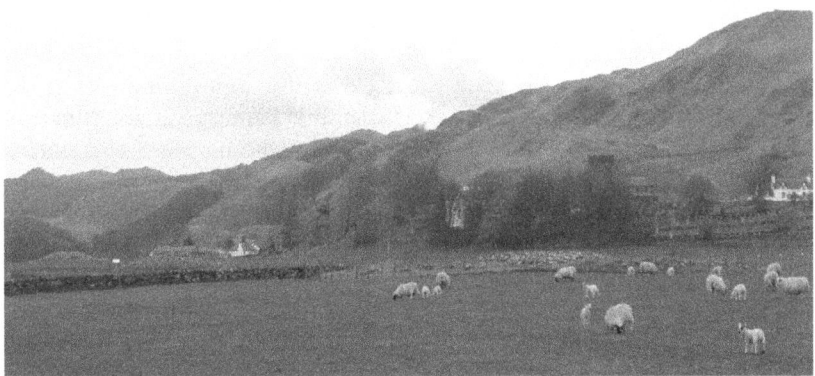

Figure 3.1 Section of the Bronze Age linear cemetery with Kilmartin Village in the background (Source: Paul Burtenshaw)

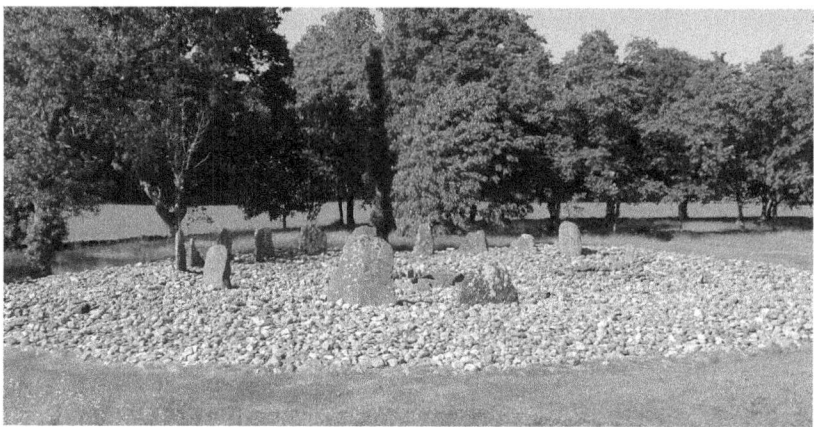

Figure 3.2 Stone circle at Templewood, Kilmartin Glen (Source: Paul Burtenshaw)

type of sites from this period, it has been described as a 'ritual landscape' (RCAHMS, 1999). There is a complex of over 20 defensive forts from the Iron Age, the most important of which is Dunadd Fort, understood to be the centre of power of the first Scots when they arrived around 500AD. There are also significant historical remains including 16th-century Carnasserie Castle, early Christian crosses and the largest selection of medieval grave slabs in Europe (KHM, 1994). The significance of Kilmartin Glen is on a world scale, and it is one of the most important archaeological areas in Scotland (KHM, 2004).

Kilmartin House Museum (the Museum) is an independent museum established in the Glen in 1997 with the aim of providing information and orientation about the archaeology. The Museum also aimed to bring economic benefits to Kilmartin and had a strong commitment to local education, as well as to the return of artefacts excavated in the Glen (KHM, 1994). The Museum charges a small entrance fee and runs a shop and café/restaurant in an adjacent building. At the time of the study, the Museum received approximately 14,000 visitors annually, while about 28,000 use the tourist services at the site over the same period. While the original business plan envisaged a self-sustaining enterprise, the Museum currently relies on a mixture of revenue from the museum and tourist services, public funding and grants. The Museum also operates a varied events calendar including craft courses and educational activities with local schools. The establishment is a focus for continued archaeological research, including excavations and surveys, as well as stewardship of the collections of artefacts from the local area.

Access to the Glen's monuments and archaeology is free and is reached via a network of dedicated carparks and paths. There are tourist information signs in the carparks, but the Museum is the only detailed source of information about the area's archaeology. Overseas visitors make up about a third of all tourists to the Glen, and the vast majority come by car as part of touring trips to Scotland (Bailie *et al.*, 2005). The archaeological monuments of the Glen received approximately 70,000 visitors per year in 2008.

In 2008, a basic economic impact assessment of tourism to Kilmartin House Museum and the archaeology of Kilmartin Glen was conducted. The research was carried out because the Museum lacked information on the economic value of the archaeology and the museum. The archaeology of the Glen has previously been, and continues to be, under threat from development and resource extraction, particularly the excavation of gravel. Knowledge of how the archaeology contributes to the local community is vital for its protection. As the Museum now partly relies on public funding, managers also felt the need to understand what economic role the establishment played in the local area.

The assessment was calculated from regional and national tourist statistics, the Museum's own accounts and interviews with local stakeholders. Economic multiplier details were available on a national level (ESU, 1993). While space does not permit a detailed review of the method, it is sufficient to say that the assessment should only be considered a rough, but robust, calculation and in this case a general guide to monetary sums involved rather than an exact tabulation. The study found that the business activities of Kilmartin House Museum had a local economic impact of over half a million pounds a year and that its activities created an additional 4–5 full-time jobs in the economy. In addition, the spending of tourists who visit the Museum on other services, such as accommodation and food, had an economic impact of about £1 million (2008 figures) within the area of a day's journey (considered the local area). Therefore, the Museum had a local economic impact of approximately £1.5 million per year, which also means that for every pound of public money the Museum receives, approximately £14.50 goes into the local economy (as public money only represents a portion of the Museum's income). Although the Museum is a so-called 'anchor attraction', these figures do not suggest that visitors come to the region only because of the museum, but they do suggest that the museum has an important economic value.

For Kilmartin Glen, the annual fiscal impact of visitors is worth £4–6.5 million in the local area, supporting around 300 full-time jobs. However, tourists visit the Glen for its natural and cultural highlights. If tourists' motivations are considered (from existing tourist surveys), it can be said that if the archaeological resources disappeared, the region would lose approximately £4 million per year. This £4 million can be considered the approximate annual economic value of the archaeology remains of Kilmartin Glen for tourism. While these numbers may reflect other attractions in the region as well, the core of the area's tourist appeal is Kilmartin Glen, which contributes considerably to local revenue and employment.

The results of this assessment demonstrate that the archaeology of Kilmartin Glen has a potential economic value that is likely supported by other attractions and services in the area. By themselves, the data support the argument for the protection of the archaeology as an economic asset, comparable with other assets in the area. However, it is crucial also to consider how the economic value and other sociocultural values intersect. The arena for this interaction of values is Kilmartin House Museum. The study demonstrates that, as well as having a cultural role, the Museum is vital as a regional economic asset. As access to the monuments is free, the Museum is the only attraction that converts visitors' interest into revenue. Due to the highly seasonal variations in tourism demand in the region, the local villages can support few dedicated year-round tourist services. As such, much of the economic benefit ends up leaking into nearby larger towns, where services have sufficient local populations to support them during off seasons. However, crucially, by providing services such as toilets, a café and shop, the Museum keeps at least some of the tourist expenditures within the Glen. Furthermore, local accommodation providers have commented that the programme of courses and cultural events staged by the Museum are vital in providing bookings outside the tourist high season. Kilmartin House Museum is not only a cultural attraction but has a clear and vital economic role in Kilmartin Glen.

The Museum's revenue is vital in providing services that also promote the sociocultural values of the local archaeology. The Museum educates visitors about the ancient monuments and history of the area, which would not be possible at the current level without tourism-generated earnings. The Museum conducts research into the monuments, which the local population is encouraged to participate in. This public archaeology activity increases the scientific value of the monuments as more information is discovered about them. The Museum carries out a detailed and well-regarded educational programme with local schools. As a result, several parents commented to the author that their children know much more about the archaeology and history of the area than they themselves ever did. Indeed, one of the Museum's founders noted that one of the chief motivations for establishing the Museum was that, as a child in the Glen, he climbed over the monuments without ever knowing what they were. The basic presence of a museum, and the fact that many important archaeological finds have been returned to it, raises the esteem of the archaeology in local people's eyes as a source of local pride and some prosperity, much as has occurred in other parts of the world (Timothy, 2015). The activities of the Museum raise public awareness of the archaeology, which attracts more tourists, thereby feeding back into the economic coffers of the community. Interestingly, the Museum also acts as an important social space for villagers. As one of the few local services open for most of the year, the café in particular serves as a social space and for events such as art exhibitions, while the Museum also houses a local children's club.

The focus of this example has been the Museum with its programmes for the public good. However, it must be kept in mind that the Museum could not exist without the revenue generated by the public interest in the local heritage. Tourism at the Glen is of a small enough scale that it does not at present appear to be causing any damage to the physical environment or influencing the local community in any significant negative way. Although the Museum does receive limited public funding, it could not fulfil its educative, conservation, scientific, or community embeddedness mandates without revenue from tourism.

Conclusion

The case of Kilmartin Glen and Museum on one level is a reminder of the mutually beneficial relationship that can exist between economic and sociocultural values through archaeology-based tourism. Although Kilmartin Glen does not yet face many of the challenges that tourism effects in other destinations, it does demonstrate the positive outcomes that can result from it. By harnessing the economic potential of public interest in archaeology, museums and other cultural institutions are able to expand other heritage values such as educating visitors and residents about the local heritage, providing social and cultural facilities, and increasing community satisfaction. As a result, these cultural institutions can promote and protect the other archaeological values, including its fiscal ones, by safeguarding the monuments for future participation.

The relationship between the values of archaeology as demonstrated here is not simply one of the economic value 'buying' the promotion of others. Educational and conservation initiatives, such as the return of displaced artefacts, have also boosted the economic value of archaeology, while purely economic endeavours, such as the museum cafés and shops, can assume a social and cultural role as well for the local community. That archaeology is a source of jobs and revenue will also enhance its importance in the eyes of local people as their own quality of life improves. The various values of cultural heritage exist in an interconnected network of interdependent relationships with one another.

Central to the discussion in this chapter is an understanding of the economic importance of both the archaeological resources themselves and their support services and infrastructures (e.g. museums). For communities whose primary, or in some cases, sole tourism resource is archaeology, the consequences of these social and economic benefits cannot be overstated.

Being able to measure and understand not just the magnitude of economic impacts, but where economic impact is felt and by whom, and how these support other values is vital in planning and managing archaeology-based heritage tourism. Quantitative analysis is rare in archaeological studies (Gould & Burtenshaw, 2014) but more should be done in order to understand archaeological economics as well as its associated values. For this sort of research to be effective, economic outcomes cannot be seen simply as an advocacy tool or the only goal of heritage industry development (Bewley & Maeer, 2014). Understanding the economic value of archaeology must feed destination planning and site management so that other heritage values can be mutually supported and benefitted. It is my hope that understanding how economic merits interplay with other values can help break down the all too frequent binary relationship between archaeology and tourism.

References

Aas, C., Ladkin, A. and Fletcher, J. (2005) Stakeholder collaboration and heritage management. *Annals of Tourism Research* 32 (1), 28–48.
Adams, J.L. (2010) Interrogating the equity principle: The rhetoric and reality of management planning for sustainable archaeological heritage tourism. *Journal of Heritage Tourism* 5 (2), 103–123.
Assefa, G. and Ambler, C. (2017) To demolish or not to demolish: Life cycle consideration of repurposing buildings. *Sustainable Cities and Society* 28, 146–153.
Australian ICOMOS (International Council on Monuments and Sites) (2013) *The Burra Charter: The Australian ICOMOS Charter for Places of Cultural Significance, 2013.* See https://australia.icomos.org/publications/charters/ (accessed 15 January 2019).
Bailie, B., Bowkett, A.M., Fiske, P. and Nichols M. (2005) Report on the Kilmartin House Museum/Kilmartin Glen survey, September 2005. Unpublished MPhil research, University of Cambridge.
Belfiore, E. (2012) 'Defensive instrumentalism' and the legacy of New Labour's cultural policies. *Cultural Trends* 21 (2), 103–111.
Bewley, B. and Maeer, G. (2014) Tourism, regeneration and the 'heritage economy'. *Public Archaeology* 13 (1–3), 240–49
Bowitz, E. and Ibenholt, K. (2009) Economic impacts of cultural heritage – Research and perspectives. *Journal of Cultural Heritage* 10, 1–8.
Brodie, N. (2014) The antiquities market: It's all in a price. *Heritage & Society* 7 (1), 32–46.
Burtenshaw, P. (2014) Mind the gap: Cultural and economic values in archaeology. *Public Archaeology* 13 (1–3), 48–58
Burtenshaw, P. and Palmer, C. (2014) Archaeology, local development and tourism – a role for international institutes. *Bulletin for the Council for British Research in the Levant* 9 (1), 21–26.
Carman, J. (2002) *Archaeology and Heritage: An Introduction*. London: Continuum.
Carman, J. (2005) Good citizens and sound economics: The trajectory of archaeology in Britain from 'heritage' to 'resource' In C. Mathers, T. Darvill and B. Little (eds) *Heritage of Value, Archaeology of Renown: Reshaping Archaeological Assessment and Significance* (pp. 43–57). Gainesville: University Press of Florida.
Carver, M. (1996) On archaeological value. *Antiquity* 70, 45–56.
Castañeda, Q.E. and Mathews, J.P. (2013) Archaeology Meccas of tourism: Exploration, protection, and exploitation. In C. Walker and N. Carr (eds) *Tourism and Archaeology: Sustainable Meeting Grounds* (pp. 37–64). Walnut Creek, CA: Left Coast Press.

Çela, A., Lankford, S. and Knowles-Lankford, J. (2009) Visitor spending and economic impacts of heritage tourism: A case study of the Silos and Smokestacks National Heritage Area. *Journal of Heritage Tourism* 4 (3), 245–256.

Coben, L.S. (2014) Sustainable preservation: Creating entrepreneurs, opportunities, and measurable results. *Public Archaeology* 13 (1–3), 278–287.

Comer, D. (ed.) (2012) *Tourism and Archaeological Heritage Management at Petra: Driver to Development of Destruction?* New York: Springer.

Darvill, T. (1995) Value systems in archaeology. In M.A. Cooper, A. Firth, J. Carman and D. Wheatley (eds) *Managing Archaeology* (pp. 40–50). London: Routledge.

Drdácký, M. and Drdácký, T. (2006) Impact of tourism on historic materials, structures and the environment: A critical overview. In R. Fort, M. Alvarez de Buergo, M. Gomez-Heras and C. Vazquez-Calvo (eds) *Heritage, Weathering and Conservation* (pp. 805–825). London: Taylor and Francis.

ESU (1993) *Scottish Tourism Multiplier Study 1992*, Volume 1. ESU Research Paper No. 31. Surrey: Surrey Research Group.

Flatman, J. (2012) Conclusion: The contemporary relevance of archaeology – Archaeology and the real world? In M. Rockman and J. Flatman (eds) *Archaeology in Society: Its Relevance in the Modern World* (pp. 291–303). London: Springer.

Global Heritage Fund (2010) *Saving Out Vanishing Heritage: Safeguarding Endangered Cultural Heritage Sites in the Developing World*. Palo Alto, CA: Global Heritage Fund.

Gould, P.G. (2017) Collision or collaboration? Archaeology encounters economic development: An introduction. In P.G. Gould and K.A. Pyburn (eds) *Collision or Collaboration: Archaeology Encounters Economic Development* (pp. 1–14). Cham, Switzerland: Springer.

Gould, P.G. (2018) *Empowering Communities through Archaeology and Heritage: The Role of Local Governance in Economic Development*. London: Bloomsbury.

Gould, P.G. and Burtenshaw, P. (2014) Guest editorial: Archaeology and economic development. *Public Archaeology* 13 (1–3), 3–9.

Graham, B., Ashworth G.J. and Tunbridge, J.E. (2000) *A Geography of Heritage: Power, Culture and Economy*. London: Arnold.

Hall, C.M. and McArthur, S. (1996) The human dimension of heritage management: Different values, different interests, different issues. In C.M. Hall and S. McArthur (eds) *Heritage Management in Australia and New Zealand* (2nd edn) (pp. 2–21). Melbourne: Oxford University Press.

Historic Scotland (2007) *Sites Guide 2007*. Edinburgh: Historic Scotland.

Hodder, I. (2010) Cultural heritage rights: From ownership and descent to justice and well-being. *Anthropological Quarterly* 83 (4), 861–882.

Holden, J. (2006) *Cultural Value and the Crisis of Legitimacy: Why Culture Needs a Democratic Mandate*. London: Demos.

ICOMOS (2017) ICOMOS Action Plan: Cultural Heritage and Localizing the SDGs. Draft version to be reviewed by ICOMOS Scientific and National Committees. See http://www.icomos.org/en/what-we-do/focus/un-sustainable-development-goals/8776-report-of-the-istanbul-meeting-on-sdgs-and-draft-action-plan-cultural-heritage-and-localizing-the-sdgs (accessed 13 January 2019).

Johnson, P. and Thomas, B. (1995) Heritage as business. In D.T. Herbert (ed.) *Heritage Tourism and Society* (pp. 170–190). London: Mansell.

KHM (Kilmartin House Museum) (1994) *Business Plan 1. August 1994*. Unpublished document. Lochgilphead, UK: Kilmartin House Museum.

KHM (Kilmartin House Museum) (2004) *Kilmartin Business Review: Final Report July 2004*. Unpublished document. Lochgilphead, UK: Kilmartin House Museum.

Klamer, A. and Zuidhof, P. (1999) The values of cultural heritage: Merging economic and cultural appraisals. In R. Mason (ed.) *Economics and Heritage Conservation: A*

Meeting Organised by the Getty Conservation Institute, December 1998 (pp. 23–61). Los Angeles: The Getty Conservation Institute.

Labadi, S. and Gould, P. (2015) Sustainable development: Heritage, community, economics. In L. Meskell (ed.) *Global Heritage: A Reader* (pp. 196–216). Oxford: Wiley-Blackwell.

Lafrenz Samuels, K. (2008) Value and significance in archaeology. *Archaeological Dialogues* 15 (1), 71–97.

Lafrenz Samuels, K. (2009) Trajectories of development: International heritage management of archaeology in the Middle East and North Africa. *Archaeologies: Journal of the World Archaeological Congress* 5 (1), 68–91.

Lipe, W. (1984) Value and meaning in cultural resources. In H. Cleere (ed.) *Approaches to the Archaeological Heritage* (pp. 1–11). Cambridge: Cambridge University Press.

Lipe, W. (2009) Archeological values and resource management. In L. Sebastian and W. Lipe (eds) *Archaeology & Cultural Resource Management: Visions for the Future* (pp. 41–63). Sante Fe: School for Advanced Research Press.

Mason, R. (ed.) (1999) *Economics and Heritage Conservation: A Meeting Organised by the Getty Conservation Institute, December 1998*. Los Angeles: The Getty Conservation Institute.

Mason, R. (2002) Assessing values in conservation planning: Methodological issues and choices. In M. de la Torre (ed.) *Assessing the Values of Cultural Heritage* (pp. 5–30). Los Angeles: The Getty Conservation Institute

Mason, R. (2008) Assessing values in conservation planning: Methodological issues and choices. In G. Fairclough, R. Harrison, J.H. Jameson and J. Schofield (eds) *The Heritage Reader* (pp. 99–124). London: Routledge.

Mason, R. and Avarami, E. (2002) Heritage values and challenges of conservation planning. In J.M. Teutonico and G. Palumbo (eds) *Management Planning for Archaeological Sites: An International Workshop Organized by the Getty Conservation Institute* (pp. 13–26). Los Angeles: The Getty Conservation Institute.

Massy, L. (2008) The antiquity art market: Between legality and illegality. *International Journal of Social Economics* 35 (10), 729–738.

McManus, R. (1997) Heritage and tourism in Ireland – an unholy alliance? *Irish Geography* 30 (2), 90–98.

Merhav, R. and Killebrew, A.E. (1998) Public exposure: For better and for worse. *Museum International* 50, 31–37.

Mourato, S. and Mazzanti, M. (2002) Economic valuation of cultural heritage: Evidence and propects. In M. de la Torre (ed.) *Assessing the Values of Cultural Heritage* (pp. 51–76). Los Angeles: The Getty Conservation Institute.

Nijkamp, P. (2012) Economic valuation of cultural heritage. In G. Licciardi and R. Amirtahmasebi (eds) *The Economics of Uniqueness: Investing in Historic City Cores and Cultural Heritage Assets for Sustainable Development* (pp. 75–106). Washington, DC: World Bank.

RCAHMS (1999) *Kilmartin Prehistoric & Early Historic Monuments: An Inventory of the Monuments Extracted from Agryll. Volume 6*. Edinburgh: The Royal Commission on the Ancient and Historical Monuments of Scotland.

Scheyvens, R. (2002) *Tourism for Development: Empowering Communities*. Harlow: Pearson.

Selvakumar, V. (2010) The use and relevance of archaeology in the post-modern world: Views from India. *World Archaeology* 42 (3), 468–480.

Silverman, H. (2002) Touring ancient times: The present and presented past in contemporary Peru. *American Anthropologist* 104 (3), 881–902.

Starr, F. (2010) The business of heritage and the private sector. In S. Labadi and C. Long (eds) *Heritage and Globalisation* (pp. 147–169). London: Routledge.

Throsby, D. (2001) *Economics and Culture*. Cambridge: Cambridge University Press

Throsby, D. (2002) Cultural capital and sustainability concepts in the economics of cultural heritage. In M. de la Torre (ed.) *Assessing the Values of Cultural Heritage* (pp. 101–117). Los Angeles: The Getty Conservation Institute.

Timothy, D.J. (2011) *Cultural Heritage and Tourism: An Introduction*. Bristol: Channel View Publications.

Timothy, D.J. (2014) Contemporary cultural heritage and tourism: Development issues and emerging Trends. *Public Archaeology* 13 (1–3), 30-47.

Timothy, D.J. (2015) Cultural heritage, tourism and socio-economic development. In R. Sharpley and D.J. Telfer (eds) *Tourism and Development: Concepts and Issues* (2nd edn, pp. 237–249). Bristol: Channel View Publications.

Timothy, D.J. and Boyd, S.W. (2003) *Heritage Tourism*. Harlow: Prentice Hall.

UNESCO (2010) *The Power of Culture for Development*. See http://unesdoc.unesco.org/images/0018/001893/189382e.pdf (accessed 14 April 2013).

VanBlarcom, B.L. and Kayahan, C. (2011) Assessing the economic impact of a UNESCO World Heritage designation. *Journal of Heritage Tourism* 6 (2), 143–164.

4 Privatization, Archaeology and Tourism

Işılay Gürsu

Introduction

Privatization is notorious for not having a clear-cut definition (Bailey, 1987; Kawashima, 1999; Ponzini, 2010; Starr, 1989). Its common use rings a bell in many people's minds and provokes a spontaneous reaction. Born out of governments' desire 'to strengthen and expand the market at the expense of the state and to increase the exposure of the public sector to market forces' (Wilding, 1990: 19), privatization is a political choice. The spread of privatization to many countries and to various sectors has resulted in a multiplication of its meanings. It is seen as an umbrella concept for a range of initiatives undertaken by the government to increase the role of the private sector in public services (Ascher, 1987: 4–7).

This chapter examines the multiple uses of the privatization concept with a focus on its implementations in the field of culture, more specifically in the field of archaeology, which is not exempt from similar pressures in other sub-fields of culture. It does so by going through the literature that examines privatization and archaeology with a special emphasis on the role that tourism plays in this blend. Tourism is a game changer in decisions related to the future of archaeology as a discipline and the archaeological heritage itself. In this regard, searching for the raison d'etre of private interventions to archaeological sites brings tourism forth as the usual suspect. The relationship between archaeology and tourism, focusing on recent concepts, as well as the ways in which the literature on privatization and archaeology portrays tourism, constitute an important part of this chapter.

Examining the ways in which privatization is conceptualized in the academic literature, in the media, and in political discourse does not guarantee a fuller understanding of its functioning on the ground. However, such a critical reading is valuable per se, since it covers a range of issues that are defined under the term privatization, from the sales of archaeological sites to charging museum admission fees in light of examples that come from different contexts such as Italy, Mexico, Greece and Cambodia. It should be noted at the outset of this chapter, however, that not all archaeology belongs to the public domain. There are many examples throughout the world of private or non-profit ownership of archaeological sites. This chapter is concerned foremost with publicly owned archaeology that changes ownership through privatization processes.

Privatization in the Cultural Sectors

Towards the end of the 20th century, privatization became one of the modus operandi of political, economic and social change. Proposed as a vigorous remedy against cumbersome and bureaucratic state mechanisms that failed to respond to the needs of modern society, it was welcomed by many. To minimize the government's role in the provision of goods and services, privatization was first introduced in countries such as the UK, the US, and some in Europe. What made it a worldwide political movement was the introduction of institutions like the International Monetary Fund and the World Bank to developing countries, as well as growing neoliberalism and a variety of other economic and political globalization processes (Dwyer & Čavlek, 2019; Wearing et al., 2019). The joining of the former communist states of Europe to change their roles in managing traditionally state-owned enterprises completed the picture (Boorsma, 1998).

Currently, there are examples of this privatization process from almost everywhere. In its simplest form, privatization can be described as 'a change in the ownership of a state-owned enterprise or service' (Köthenbürger et al., 2006: x). Although this definition would still correspond to many people's understanding of the concept, as the range of enterprises and services subject to privatization initiatives expands, it fails to cover other activities that are increasingly associated with privatization. Privatization, as well as its detailed discussions by scholars, can be observed in many different sectors that have been traditionally attributed to the public realm, such as health care (Chen, 2013; Toebes et al., 2014), education (Kishan, 2008; Murphy et al., 1998; Pring, 1987), culture (Boorsma, 1998; Wu, 2005; Yúdice, 1999), postal services (Parker & Saal, 2003) and even prisons (Feeley, 1997).

Although methods in each sector differ, Wu (2005) lists four new steps that come in the same package with privatization and which characterize the change in state ideology: decreasing the value of the state, abolishing government intervention, total privatization and management culture. In a nutshell, for those in favour of privatization, these four steps are the signifiers of efficient use of public funds, access to more professional services and financial surplus. The opposing minds to privatization, on the other hand, criticize it due to its potential exploitation of public resources to create profits for a limited group. In some cases, because the rationales and justifications are based on ideologies, privatization has been uncritically applied to several sectors and countries (Starr, 1989).

As a result of these changes in state ideologies and pressures arising due to a free market economy, private intervention in the field of culture has become commonplace. Vogelsang (2006) classifies definitions of privatization in relation to the cultural sector under three main categories. He distinguishes simple privatization, which is a change in ownership, from liberalization and deregulation. Liberalization refers to the changes in the rules of the market, market participation and conduct, whereas deregulation is the change in public regulation and the introduction of fewer constraints to the market. Deregulation refers to the increased autonomy that puts organizations at a greater distance from the government.

Related to the concept of autonomy, Boorsma (1998) gives five other subcategories. The first is the sale of public organizations or public assets; the second one is the creation of a more internally independent public or private organization. The third is contracting out, and the fourth refers to the mobilization of volunteers through which non-public workers carry out work for the public. The last concept of autonomy is the application of user fees. In this sense, privatization refers to a change in ownership, the legal status of the organization, the type of personnel doing the work and most importantly, the funding. Another distinction in the privatization argument seems to be the one between financing, which refers to investment and funding, and sales of public belongings. This is an important point since the upcoming parts of this chapter touch upon the debates about the privatization of archaeology. Whenever there is any reference to the sale of archaeological property, discussions are invariably heated, and all the scholars who write about it are against it. When the argument regards finances, however, there are differing views.

Boorsma (1998) laments that articles on privatization are generally confined to a specific type of privatization, failing to come up with a wider framework, and therefore he notes seven different types of privatization in the cultural sector:

- divestiture: the organization is sold to a private enterprise;
- free transfer of property rights;
- transformation of a state organization into a more independent organization such as foundations or trusts;
- the agency model, which empowers a public manager in the course of 'new managerialism' and refers to self-administered integral management;
- contracting-out in which work is done by hired private companies, including security and cleaning;
- use of volunteers
- private funding.

Reliance on the private sector for providing the services that traditionally belong to the state is favoured by an increasing number of countries around the world. However even in those countries, when cultural or archaeological heritage becomes the subject of privatization, the support, if any, is always conditional. For instance, Peacock (1995: 192) touches upon the subject of privatization within the framework of supply of heritage services with some reservations. He states, 'heritage services without pure public goods characteristics could be privatized in one form or the other, but with activities regulated and possibly subsidized to conform to heritage objectives'. In line with Boorsma's (1998) classifications, Peacock (1995) mentions the application of user charges in museums or putting publicly operating services out to competitive tendering as possible privatization options. Still, he argues that privatization does not necessarily require heritage services to be provided by for-profit enterprises. One of the main reasons why these enterprises should not, or could not, take the place of the state for providing heritage services is underlined by Canclini (2001). He states, 'in some countries, the cultural action of the public sector was reduced to protecting the historical heritage (museums, archaeological sites, etc.) and promoting traditional arts (visual arts, music, theatre, literature). The premise here is that, given declining attendance, these forms of culture would not survive without artificial respiration from the government' (Canclini, 2001: 100).

Apparently, in the field of archaeology, there are various transactions that can be related to privatization, but the level of this association may differ in scale. In other words, contracting out the management of a cafe at an archaeological site is not equal to leasing the total site to a private party in terms of its association with privatization. As Schuster (1998) suggests, in these cases it is difficult to pin down the tipping point at which privatization happens in any systematic way, and there would be many disagreements on the exact moment of the occurrence of the tipping point, but many would agree that one exists in every place.

New Perspectives in Archaeology and its Relationship with Tourism

In 2018, there was a new exhibition in Istanbul titled 'The Curious Case of Çatalhöyük', developed to celebrate the 25th year of excavations at the world famous site.[1] Giving an account of the scientific methodologies, research questions, as well as the glimpses of how life would have been in the Neolithic town of Çatalhöyük, the exhibition touches upon many aspects of archaeology. One thing worth underlining for the purposes of this chapter is that the exhibition makes it possible for the interested public to communicate with the excavation director, Ian Hodder. Upon collecting questions from the curious public through social media, the organizers forwarded them to Professor Hodder. His answers were available for exhibition visitors. This is a very good example of the new approaches to archaeology. The Çatalhöyük excavation is an exceptional case and does not yet represent a common approach; however, it is setting new standards for archaeology. It is living proof of archaeologists coming out of their fortresses, as described by Ling (2015), not only because there is peer pressure but also because of the growing number of people in the general public who are interested in their work and who are demanding answers to their questions.

While the definition of archaeology as a scientific discipline is satisfactory, the post-processual approach to archaeology has resulted in an increase in the tasks that archaeologists are expected to deal with to make archaeology more relevant in the modern world (Rockman & Flatman, 2013). This has resulted in the appearance of different disciplines and sub-disciplines. Increasing demand for better management, conservation, and presentation gave rise to cultural heritage management. Putting people at the forefront of all these initiatives brought about the notion of public archaeology. Questioning for whom archaeology is being practiced, and placing local communities at the centre, has become the main concern for archaeological ethnographies and community archaeology.

In 1979, Mexican archaeologists Rebeca Panameño and Enrique Nalda asked 'Arqueología para quién?' (Archaeology for whom?). These new approaches demonstrate that the question is still valid and remains a radical consideration for the discipline (McGuire, 2007). Castañeda and Mathews' (2013: 45) argument on the assignment of the proper heirs to archaeological heritage by archaeologists provides interesting insight on the 'archaeology for whom' question. They argue that this agenda 'runs the risk of imposing an artificial conception of proper ownership if the local stakeholders have not been or are not yet properly interpellated as proper heirs of archaeological stewardship'. On top of this, 'trouble may arise for archaeologists, however, when the descendent-stakeholders refuse to be contained by the archaeologists' conception of what it is to be a good citizen-heir of archaeological heritage' (Castañeda & Mathews, 2013: 56). Castañeda and Mathews base their arguments on the case of Mexico, but their arguments can shed light on other countries' experiences as well, albeit with slight differences.

For instance, in Middle Eastern countries where there is often a conflictual relationship between archaeological heritage and contemporary society, the problem is with the notion of descendent-stakeholders. In some cases, the community refuses to acknowledge rather than claim to be related to the past civilizations whose ruins lay around them. However, this does not mean that they would be indifferent to whatever happens to their surroundings. There are a couple of reasons behind this. The first, and most direct, is linked with the economic potential of the site, mostly through tourism (Ghanem & Saad, 2015; Tarawneh & Wray, 2017). Although a high number of paying visitors to an archaeological site does not guarantee an equal distribution of revenue within the local community, it is still a lucrative business. To eliminate this unequal distribution, the 8th Draft of the International Cultural Tourism Charter, Managing Tourism at Places of Heritage Significance was adopted by ICOMOS in 1999. It strongly emphasizes the role and well-being of the local community. The fifth principle of the charter states that tourism and conservation activities should benefit the host community (ICOMOS, 1999).

As Walker and Carr (2013: 27) put it, 'archaeologists are increasingly aware of the importance of public awareness and understanding of their work, not only out of kindness but also out of sustaining funding for their work'. While this notion holds true for western contexts where political entities, as the main funding bodies for archaeological work, are held accountable for their decisions and therefore take public opinion seriously, it is not always valid in non-western contexts where politics do not always have the same transparency and accountability constraints. In these cases, archaeological heritage has a greater chance of being protected, as long as it has economic value, mostly through tourism.

The relationship between archaeological heritage and tourism has been discussed extensively (Adams, 2010; Boniface & Fowler, 1993; Staiff *et al.*, 2013; Timothy & Boyd, 2003, 2006; Walker & Carr, 2013). As an historical anecdote, Timothy (2011: 341) states that 'much of the world's knowledge of the human past has come to light through the labours of amateur archaeologists or of volunteers undertaking archaeology experiences during their vacations, especially in the early years of archaeological exploration'. This is still true today. Besides volunteer tourists at archaeological sites, contemporary archaeologists are some of the most enthusiastic tourists to explore other sites that are similar to theirs, sometimes out of professional interest but also out of pure curiosity, which might have been the driving force for their selection of this profession.

Additionally, as Castañeda and Mathews (2013: 47) put forward, 'archaeology has become increasingly connected to tourism in terms of popular imaginaries and representations of the past, if not also economically dependent upon it for the production of the ideological appreciation of the past'. One of the main conflicts between tourism and archaeology is grounded in the difference that is contained in the core of these two activities. Tourism is an industry. It is well adapted to keeping an eye on its market, equipped with tools to measure economic success and take action when there are changes in consumer demand. There are some attempts to shape these changes, but more often tourism follows the trends; it concentrates on the demand side.

Archaeology, even when its boundaries expand beyond scientific fieldwork, is by nature more static. Archaeological projects are often lifetime projects for many directors. Therefore, the decisions taken about the site rarely concentrate on short-term adaptations. In economic terms, they are more concentrated on the supply side. Therefore, following Slick's (2002) suggestion to consider tourism a partner with archaeology rather than an adversary is achievable only to some extent. The gap is inevitably getting smaller, not out of a more congenial atmosphere, but as a result of tourists demanding access to first-hand and scientific information about the places they visit. Archaeologists, therefore, become a source to turn to. On the other hand, archaeology is not exempt from the market pressures that other cultural sectors face. Archaeologists need to communicate the importance of their work as a means of sustaining funding for their excavations, which are costly. This is one of the reasons privatization trends are on the rise in the context of archaeology.

Privatization in Archaeology and the Role of Tourism: Global Perspectives

From Italy to Mexico, from Cambodia to Greece, many authors have studied, criticized or promoted privatization as a government strategy in relation to archaeological heritage. Thanks to this academic debate, it is possible to look at the multiple uses of privatization in relation to archaeology.

Academicians are not the only ones interested in this topic. The media and politicians have also commented extensively. By looking at these sources, this section aims to show the widespread use of privatization to describe recent developments in the world of archaeology. I also look at the various ways of integrating tourism in this discussion, rather than simply concluding that tourism is always the usual suspect in decisions favouring the insertion of private actors into the cultural sector. On the contrary, there are some cases in which tourism is used as a discourse to strengthen the arguments against privatization.

The first example regards Italian cultural heritage, with a special emphasis on archaeological sites. In his multi-faceted examination of the privatization attempts that have materialized due to changes in Italian law, Benedikter (2004) defines privatization as a change in the ownership of Italian archaeological property. His concerns are grounded on the proposal of a new act that made it possible to sell property or land with cultural heritage status to private investors. He also argues that privatization in the form of sponsorship could have been an acceptable action over selling (Benedikter, 2004). Neoliberalism, bureaucratic difficulties, and costs related to cultural heritage are listed as the motivations behind this attempt. The minister at the time, Giulio Urbani, described the situation in a remarkable way: 'Italy is like a person with many houses, but also with many debts. So we have to look at which houses are dispensable' (quoted in Benedikter, 2004: 370).

This populist analogy makes no reference to the qualities of these imaginary houses. Were they inherited from beloved uncles, aunts or parents? Which ones will be sold for the sake of the others? Will I still be able to visit them if I want to remember my beloved aunt, or should I forget about it because it will become someone else's private home? Even though without a right to access, can I be sure that these houses will continue to exist at least, or will they be subject to new development opportunities?

Opponents of this law have raised similar questions. For instance, on the topic of privatizing state-owned cultural heritage, Palumbo (2006) is quite critical. Defining cultural heritage exploitation and cultural heritage use as two different methods of heritage management, he strongly argues that one of the key issues is to control the quality of private intervention. He does not believe that the private sector should be kept out of the picture completely; however, he does not hesitate to borrow Settis' (2007) idea that improving services with the help of the private sector is one thing; expecting the private sector to support conservation and maintenance is another.

One important point forwarded in Benedikter's (2004) article regards the role of tourism in privatization decisions. As a reaction to the inclusion of Alba Fucens, an archaeological site in Abruzzo, among the cultural property that could be sold, the mayor states: 'Alba Fucens must remain public property so that it can continue to be an important tourism destination' (quoted in Benedikter, 2004: 378). The mayor thus appears to suggest that privatization would spell an end to tourism.

Another investigation into the process of privatizing cultural heritage, including archaeological remains, is that of Ponzini (2010). Concentrating on Italy and with reference to public administration, he defines it as 'the introduction of private actors, objectives, and modes of action' (Ponzini, 2010: 508). It can also mean a type of partnership including tourism since most tourism is also for private purposes. Analysing the transformation of institutional organizations, he lists three types of privatization. The first one refers to the alienation and securitization of the state-owned historic real estate in question. The second type regards the establishment of mixed public-private entities to manage and promote cultural heritage. Third is the introduction of private actors into policy making and implementation (Ponzini, 2010). The underlying motivation for privatization is the concerns over how public expenditures should be allocated.

An example from Mexico touches upon very interesting points. Breglia (2006) defines privatization as property ownership, the establishment of private economic enterprises within the territory of the state and state privatization of federal resources. In a detailed historical account of the privatization of Chichen Itza, a world-famous complex of Maya ruins on the Yucatán Peninsula, Breglia (2006) emphasizes two episodes in the site's life cycle. The earlier is concerned with the private ownership of a national archaeological site with many references to colonialism. She describes the effects of private ownership on land use and subsequently the local communities. The latter she defines as 'neo-liberal privatization', or the transformation from 'jungle-covered ruins to a renowned international tourist destination' (Breglia, 2006: 66).

While the site continues to be privately owned, the earlier owners were much more interested in excavating it and exporting the archaeological materials abroad, whereas the more recent owners were tourism entrepreneurs who built hotels on the site and opened it for mass tourism, the revenue from which is not shared with the local community. While the concern at Alba Fucens, Italy, was that private ownership would hinder the ruins' potential use for tourism, the example of Chichen Itza shows a different perspective. In the Mexican case, tourism thrives at Chichen Itza, but its benefits for nearby communities and care for the site itself have taken a back burner to the needs of tourism.

Kreutzer (2006) also writes about two cases in Mexico. In his first example, he focuses on privatization in terms of its association with the ownership issue and portrays a potentially successful public archaeology project that was undertaken on private land with cooperation by the owner. However, when the real estate changed hands, the new owner stopped the excavations and the project ended suddenly. The end result of this case of private ownership was restricted access to the site; neither archaeologists nor tourists are welcome. The second example also concerns land use. The difference, however, regards the construction of a Walmart superstore in close proximity to an archaeological site. Although tourism is one source of income for the community, the job opportunities created by the superstore are also regarded as lucrative. According to Kreutzer (2006: 61), this has negative impacts on the preservation of the tangible heritage as 'economic motivations override concern for preserving the past'. Kreutzer (2006: 63) also notes that local shop owners, who are directly affected by the mega Walmart say: 'the ruins and us go together'.

An example from Cambodia concentrates on the private ownership of artefacts and links the discourses of colonialism and commodification. According to Anderson (2007), there is a direct link between tourism and looting. As early as 1908, curators found themselves having to prevent visitors from looting Angkor Wat. The 1990s tourism boom also resulted in increased plundering. Owing to corruption and tight government control of planning and development decisions, which favour international corporations and government officials, local inhabitants are excluded from the benefits of heritage tourism (Anderson, 2007). At Angkor Wat, privatization is used to justify or explain residents being precluded from the economic benefits of tourism and the illegal trafficking of ancient relics.

Although not always linked directly with archaeology, privatization is also common in museum settings. Studies reveal that there are differing views of museum privatization. For instance, Engelsman (1996) examines privatization in Dutch museums and claims it to be a huge success for the museums. He suggests that privatization might be a misleading term for the process the museums underwent. However, the process aligns with Boorsma's (1998) assertions noted earlier. Many museums have become private-sector organizations tasked with caring for, exhibiting and studying national collections – duties that devolved from former government responsibilities. In the Netherlands, public funds continued to be available and even increased during the privatization process. One of the greatest outcomes of this process was that museums were better connected with the needs of society (Engelsman, 1996).

Kawashima (1999) examines the privatization of museum services in UK local authorities from three different perspectives. The first is the change in legal status of the museum to a charitable trust or non-profit organization. The second is contracting in or buying in private service providers. The last perspective is the establishment of a market and marketing orientation. One interesting point concerns the relationship between tourism and privatization: that municipalities have a tendency to take over bankrupt private museums because the acquisition is seen as adding to the cities' tourism resource base (Kawashima, 1999).

Another study of museums and privatization in Taiwan looks at the impact of privatization on museum admission charges (Chung, 2005). Chung approaches the concept of privatization with caution and sets the framework as the transfer of authority over collections and finding alternative funding sources besides the state. In this regard, referring to Schuster's (1998) hybridization concept, Chung (2005) compares two museums' pricing strategies, as well as the ways in which newly available funds are used. According to his study, the museum with a stronger tourism orientation was more inclined to raise its entrance fees.

Another critical approach to the incursion of private actors and interests in archaeology can be seen in India. Stein (2011) analyses the aesthetic changes that emerge as a result of tourism's use of temples and links this argument to a private scheme: Adopt-a-Monument. In this scheme, a private entity assumes financial responsibility for a given monument, including its conservation, restoration and maintenance.

Tschmuck (2006) discusses the Austrian experience in privatizing cultural heritage, which did not involve sales to private parties but transferred control to organizations that were less dependent on the state. The term *Ausgliederung* is used to describe a model in which organizations receive public or private legal status but remain entirely in public ownership. The rationale behind this change is increased managerial autonomy for the museums, reduced bureaucracy, and improved efficiency without additional budget burdens. This transformation signals the 'commercial' approach of *Ausgliederung* in the sense that the privatized institutions were expected to respond to market demands, including tourism (Tschmuck, 2006).

Disputes are not only the domain of scholarly literature. Whenever there is a decision to introduce private interest to the field of archaeology, a media debate usually follows. For instance, *Time* magazine's article titled 'Can Privatization Save Greece?' sparked a lot of discussion on the subject (McDonald-Gibson, 2014). An archaeologist working in Greece recently suggested letting private companies take over the development, promotion, and security of certain under-exploited sites in exchange for a share of the revenue generated from tourists. This proposal was met with resistance by the Society of Greek Archaeologists. 'Archaeologists and scholars and the entire personnel of the Archaeological Service will never let anyone covet the management of archaeological sites and monuments of our country' (Archaeology News Room, 2014: n.p.).

A similar discussion broke out when the following headline appeared: 'Private Companies to Manage Peru's Archaeological Sites' with regard to a law that would allow private companies to provide tourist services at heritage sites (Post, 2015a). In the Greek example, there was no legal arrangement, only a suggestion by an archaeologist that privatization might be a remedy to managing the sites better. In Peru, the concept was introduced as 'partial privatization', which means that ownership of the sites would remain with the state, whereas the tourist services might be outsourced. In less than a month, this law was repealed owing to protests, especially in Cusco, the former Inca capital and home of the Machu Picchu ruins (Post, 2015b).

An example from Turkey also shows how privatization is promoted by politicians. Turkey, like France and other continental European countries, has a centralized approach to managing the use and ownership of cultural properties. In the last 15 years, there have been many legal changes regarding the use and preservation of cultural assets. As a result, new approaches to managing Turkey's archaeological assets have appeared to implement more flexible methods. It was the first time partnerships with the private sector were officially considered a viable option. In August 2005, during a Turkish Ministry of Culture and Tourism meeting, the minister at the time, Atilla Koç, announced that the management of museums and archaeological sites would be privatized:

> The first examples chosen for the privatization are the management of Topkapı Palace and St. Sophia. This is the first initiative of its kind in Turkey and we have worked very well on the regulations and we decided to try it on some sites. If we can adopt this system to all of our archaeological sites and have Turkish companies involved in this business we will make great progress regarding our income and the preservation of our sites. If we have private firms, their performance based evaluations will put an end to the scandals in the sector. (Atilla Koç, quoted in Ekmekçi, 2006)

Conclusion

Renfrew and Bahn (1991) define archaeology as the study of the human past. However, this definition does not cover all archaeological activities that comprise the discipline as practiced in the 21st century. The boundaries of archaeological work are being enlarged. For example, expectations from an archaeological excavation have shifted from solely a scientific work to a more community-based endeavour that involves education, conservation of ancient ruins, and public outreach. Therefore, the movement toward 'private intervention in archaeology' refers not only to the situation where the costs of an excavation and investigation are covered by a private company but also includes all other activities alongside the scientific work becoming a source of income beyond state coffers. These activities constitute an important part of the practice of archaeology and include ownership, renting and acquiring archaeological property, managing archaeological sites and sponsoring archaeological practices. Different modes of privatization are evident in these areas of practice, and tourism can be a decisive factor for the configuration of these modes.

This chapter has shown that private ownership of archaeological property is met with a lot of resistance. In these cases, tourism is usually at the centre of these discourses owing to its potential to create jobs and earnings for the broader community, and it is often used as a discourse against privatization. The reactions are varied, however, regarding the management or sponsorship of archaeology, especially when it comes to finances. Some refer to the success of outsourcing and greater autonomy from the state. In these instances, tourism is criticized by those who are against bringing archaeology into the private sector. Some observers and critics are concerned about losing authenticity in the transfer of archaeology from public to private hands. This suggests that perhaps public ownership is seen as more protective of artefacts' objective authenticity, while in private hands, archaeology might become commoditized and under the domain 'subjective authenticity', which has been more closely connected to tourism (Chhabra, 2012).

To conclude, tourism is an important component of the framework of public archaeology. As a strong supporter of this approach, I conditionally welcome archaeology's increasing involvement with tourism and vice versa, which has clear links to privatization. As long as the economic and social benefits of tourism for the community are not compromised for the sake of generating revenue for a few companies and elite local leaders, and as long as the involvement of community members is ensured, tourism and private intervention in archaeology can become an engine for growth and long-term and sustainable archaeological practice. Many good examples, such as the work of the Sustainable Preservation Initiative,[2] show how this can be successful on the ground. Making archaeology more relevant to contemporary society causes archaeologists to reconsider the multiple meanings of tourism and privatization; this call is too important to ignore for archaeology professionals in the 21st century.

Notes

(1) Hosted at Koç University, Anatolian Civilizations Research Center, ANAMED (June 2017–February 2018): https://anamed.ku.edu.tr/en/curious-case-catalhoyuk

(2) http://www.sustainablepreservation.org. 'The Sustainable Preservation Initiative creates economic opportunities by giving communities the tools to be self-reliant, leveraging their historic sites responsibly and freeing them to thrive'

References

Adams, J.L. (2010) Interrogating the equity principle: The rhetoric and reality of management planning for sustainable archaeological heritage tourism. *Journal of Heritage Tourism* 5 (2), 103–123.

Anderson, W. (2007) Commodifying culture: Ownership of Cambodia's archaeological heritage. *Limina: A Journal of Cultural and Historical Studies* 13, 103–112.

Archaeology News Room (2014) Privatizing cultural heritage in Greece, Archaeology and arts, *wiki blog*, 21 January 2014. See https://www.archaeology.wiki/blog/2014/01/21/privatizing-cultural-heritage-in-greece/ (accessed 30 May 2015).

Ascher, K. (1987) *The Politics of Privatisation*. Basingstoke: Macmillan.

Bailey, R.W. (1987) Uses and misuses of privatization. *Proceedings of the Academy of Political Science* 36 (3), 138–152.

Benedikter, R. (2004) Privatisation of Italian cultural heritage. *International Journal of Heritage Studies* 10 (4), 369–389.

Boniface, P. and Fowler, P. (1993) *Heritage and Tourism in the Global Village*. London: Routledge.

Boorsma, P.B. (1998) Privatizing the muse 'and all that jazz'. In P.B. Boorsma, A. van Hemel and N. van der Wielen (eds) *Privatization and Culture: Experiences in the Arts, Heritage and Cultural Industries in Europe* (pp. 23–45). Dordrecht: Kluwer.

Breglia, L.C. (2006) *Monumental Ambivalence: The Politics of Heritage*. Austin: University of Texas Press.

Canclini, N. (2001) *Consumers and Citizens: Globalization and Multicultural Conflicts*. Minneapolis: University of Minnesota Press.

Castañeda, Q. and Mathews, J. (2013) Archaeology Meccas of tourism: Exploration, protection and exploitation. In C. Walker and N. Carr (eds) *Tourism and Archaeology: Sustainable Meeting Grounds* (pp. 37–64). Walnut Creek, CA: Left Coast Press.

Chhabra, D. (2012) Authenticity of the objectively authentic. *Annals of Tourism Research*, 39 (1), 499–502.

Chen, W. (2013) Does healthcare financing converge? Evidence from eight OECD countries. *International Journal of Health Care Finance and Economics* 13 (3/4), 279–300.

Chung, Y. (2005) The impact of privatization on museum admission charges: A case study from Taiwan. *Marketing Management* 7 (2), 27–39.

Dwyer, L. and Čavlek, N. (2019) Economic globalization and tourism. In D.J. Timothy (ed.) *Handbook of Globalization and Tourism* (pp. 12–26). Cheltenham: Edward Elgar.

Ekmekçi, H. (2006) Yeni Şafak'ta Şafak Attı mı? *Milli Çözüm Dergisi*. See http://millicozum.com/mc/kasim-2005/yeni-safakta-safak-atti-mi (accessed 30 May 2015).

Engelsman, S. (1996) Dutch national museums go 'private'. *Museum International* 48 (4), 49–53.

Feeley, M. (1997) The privatization of prisons in historical perspective. In M. McShane (ed.) *The Philosophy and Practice of Corrections* (pp. 133–143). London: Routledge.

Ghanem, M.M. and Saad, S.K. (2015) Enhancing sustainable heritage tourism in Egypt: Challenges and framework of action. *Journal of Heritage Tourism* 10 (4), 357–377.

ICOMOS (1999) *International Cultural Tourism Charter*. See https://www.icomos.org/charters/tourism_e.pdf (accessed 20 June 2018).

Kawashima, N. (1999) Privatizing museum services in UK local authorities. *Public Management: An International Journal of Research and Theory* 1 (2), 157–178.

Kishan, R. (2008) *Privatization of Education*. New Delhi: APH Publishing Corporation.
Köthenbürger, M., Sinn, H. and Whalley, J. (2006) Introduction. In M. Köthenbürger, H. Sinn and J. Whalley (eds) *Privatization Experiences in the European Union* (pp. ix–xviii). Cambridge, MA: MIT Press.
Kreutzer, D. (2006) Privatising the public past: The economics of archaeological heritage management. *Archaeologies* 2 (2), 52–66.
Ling, L. (2015) Archaeology: Sharing with whom? A review of 'Excavation Report of Hezhang Kele Site in 2000'. In P. Stone and Z. Hui (eds) *Sharing Archaeology: Academe, Practice and the Public* (pp. 47–56). London: Routledge.
McDonald-Gibson, C. (2014) Can privatization save the treasures of ancient Greece? *Time*, January 18 2014. See http://world.time.com/2014/01/18/can-privatization-save-the-treasures-of-ancient-greece/ (accessed 30 May 2015).
McGuire, R.H. (2007) Foreword: Politics is a dirty word, but then archaeology is a dirty business. In Y. Hamilakis and P. Duke (eds) *Archaeology and Capitalism: From Ethics to Politics* (pp. 9–10). Walnut Creek, CA: Left Coast Press.
Murphy, J., Gilmer, S., Weise, R. and Page, A. (1998) *Pathways to Privatization to Education*. Greenwich: Ablex Publishing Corporation.
Palumbo, G. (2006) Privatization of state owned cultural heritage: A critique of recent trends in Europe. In N. Agnew and J. Bridgland (eds) *Of the Past, For the Future: Integrating Archaeology and Conservation* (pp. 35–39). Los Angeles: Getty Conservation Institute.
Panameño, R. and Nalda, E. (1979) Arqueología para quién? *Nueva Antropología* 12, 111–124.
Parker, D. and Saal, D. (eds) (2003) *International Handbook on Privatization*. Cheltenham: Edward Elgar.
Peacock, A.T. (1995) A future for the past: The political economy of heritage. *Proceedings of the British Academy* 87, 189–243.
Ponzini, D. (2010) The process of privatisation of cultural heritage and the arts in Italy: Analysis and perspectives. *International Journal of Heritage Studies* 16 (6), 508–521.
Post, C. (2015a) Private companies to manage Peru's archaeological sites. *Peru Reports*, 28 September 2015. See https://perureports.com/2015/09/28/private-companies-to-manage-perus-archaeological-sites/ (accessed 20 December 2016).
Post, C. (2015b) Peru repeals tourism privatization law after protests. *Peru Reports*, article, 23 October 2015. See https://perureports.com/2015/10/23/peru-repeals-tourism-privatization-law-after-protests/ (accessed 20 December 2016).
Pring, R. (1987) Privatization in education. *Journal of Education Policy* 2 (4), 289–299.
Renfrew, C. and Bahn, P.G. (1991) *Archaeology: Theories, Methods, and Practice*. New York: Thames and Hudson.
Rockman, M. and Flatman, J. (eds) (2013) *Archaeology in Society: Its Relevance in the Modern World*. New York: Springer.
Schuster, M.J. (1998) Beyond privatization: The hybridization of museums and the built heritage. In P.B. Boorsma, A. van Hemel and N. van der Wielen (eds) *Privatization and Culture: Experiences in the Arts, Heritage and Cultural Industries* (pp. 58–81). Dordrecht: Kluwer Academic Publishers.
Settis, S. (2007) *Italia S.p.A. L'assalto al patriminio culturale*. Torino: Piccola Biblioteca Einaudi.
Slick, K. (2002) Archaeology and the tourism train. In B. Little (ed.) *Public Benefits of Archaeology* (pp. 219–227). Gainesville, FL: University Press of Florida.
Staiff, R., Bushell, R. and Watson, S. (eds) (2013) *Heritage and Tourism: Place, Encounter, Engagement*. London: Routledge.
Starr, P. (1989) The meaning of privatization. In S.B. Kamerman and A.J. Kahn (eds) *Privatization and the Welfare State* (pp. 15–48). Princeton: Princeton University Press.

Stein, D. (2011) To curate in the field: Archaeological privatization and the aesthetic 'legislation' of antiquity in India. *Contemporary South Asia* 19 (1), 25–47.

Tarawneh, M.B. and Wray, M. (2017) Incorporating Neolithic villages at Petra, Jordan: An integrated approach to sustainable tourism. *Journal of Heritage Tourism* 12 (2), 155–171.

Timothy, D.J. (2011) *Cultural Heritage and Tourism: An Introduction*. Bristol: Channel View Publications.

Timothy, D.J. and Boyd, S.W. (2003) *Heritage Tourism*. Harlow: Prentice Hall.

Timothy, D.J. and Boyd, S.W. (2006) Heritage tourism in the 21st century: Valued traditions and new perspectives. *Journal of Heritage Tourism* 1 (1), 1–16.

Toebes, B., Ferguson, R., Markovic, M. and Nnamuchi, O. (2014) *The Right to Health: A multi-country Study of Law, Policy and Practice*. Dordrecht: Springer.

Tschmuck, P. (2006) The budgetary effects of 'privatizing' major cultural institutions in Austria. *The Journal of Arts Management, Law and Society* 35 (4), 293–304.

Vogelsang, I. (2006) Network utilities in the United States: Sector reforms without privatization. In M. Köthenbürger, H. Sinn and J. Whalley (eds) *Privatization Experiences in the European Union* (pp. 50–85). Cambridge, MA: MIT Press.

Walker, C. and Carr, N. (2013) Tourism and archaeology: An introduction. In C. Walker and N. Carr (eds) *Tourism and Archaeology: Sustainable Meeting Grounds* (pp. 11–36). Walnut Creek, CA: Left Coast Press.

Wearing, S., McDonald, M., Taylor, G. and Ronen, T. (2019) Neoliberalism and global tourism. In D.J. Timothy (ed.) *Handbook of Globalization and Tourism* (pp. 27–43). Cheltenham: Edward Elgar

Wilding, P. (1990) Privatisation: An introduction and a critique. In R. Parry (ed.) *Privatization* (pp. 18–31). London: Jessica Kingsley Publishers.

Wu, C. (2005) *Kültürün Özelleştirilmesi: 1980'ler Sonrasında Şirketlerin Sanata Müdahalesi*. Istanbul: İletişim.

Yúdice, G. (1999) The privatization of culture. *Social Text* 59, 17–34.

5 Marketing Archaeological Heritage for Tourism

Alan Fyall, Anna Leask and Sarah B. Barber

Introduction

As evident throughout this chapter, archaeological tourism helps attract new markets to destinations and helps differentiate destinations in highly competitive markets. In addition, the specific location and historical context of archaeological attractions serve as a means to enhance the authentication of destinations, with noted contributions to place identity and richness of the visitor experience. From a marketing perspective, this chapter introduces the domain of cultural and heritage tourism and the many benefits it can bring to tourist destinations before introducing the more specific contribution that archaeology and archaeological sites bring to tourism as branding mechanisms and foci of marketing campaigns.

After elaborating on marketing issues in the context of archaeology, this chapter examines two empirical examples to illustrate particular marketing facets of archaeological tourism. The first example provides insight into cultural tourism development in Oaxaca, Mexico, home to two of Mexico's UNESCO World Heritage Sites: the ancient city of Monte Albán and the cave networks of Yagul and Mitla (UNESCO, 2017). Beyond the state capital, Oaxaca City, far less attention has been given to archaeological resources along the state's 597 km of Pacific coastline. This case outlines how archaeological resources provide a suitable means for tourism development and marketing since they contribute significantly to destination differentiation from the many competing sun, sea and sand (SSS) beach destinations in the Caribbean, this especially being the case for visitors from North America. The second example draws attention to the recent 'History Bug' marketing campaign launched by Historic Environment Scotland. This imaginative sales-driven concept was designed to capitalize on Scotland's Year of History Heritage and Archaeology, with 2017 being the best year to 'catch the history bug', with membership and attendance at events the only cure! These two cases help highlight many of the benefits of archaeological tourism. The chapter concludes with some valuable thoughts for the future development and marketing of this niche form of tourism.

Cultural and Heritage Tourism

As noted in the introductory chapter of the book, archaeology is an important part of the broader domain of heritage. The marketing principles elucidated in this chapter are equally applicable to archaeology as they are to other manifestations of cultural heritage. Cultural or heritage tourists can be classified as people who visit cultural attractions or events as part of their trip. For many, these visits are a significant part of the journey, with the performing arts, visual arts and crafts, festivals, archaeological sites, museums and cultural centres, historic sites and interpretive centres all included under the wider umbrella of cultural and heritage tourism (Canadian Tourism Commission, 1999). The Pacific Asia Tourism Association (PATA) estimates that cultural and heritage tourism is growing at a rate of 15% year-on-year, with an estimated 37% of all international travellers including a cultural component in their trip (PATA, 2015). In addition, the World Tourism Organization (UNWTO) indicates that more than 50% of global tourism is motivated, in part, by a desire to experience culture and heritage with the direct global value of heritage tourism estimated at well over US $1 billion. Cultural heritage-based tourism is also believed to have a yield, in that cultural tourists on average spend more per day and stay longer than other tourists with higher and growing levels of repeat visitation in many markets.

If one were to interpret cultural and heritage tourists strictly as those who are motivated purely to visit a cultural attraction, then the market size is estimated at between 5 and 8%. Heritage travellers tend to be in the 39–59 age bracket, are better educated, have broader travel experiences, are more quality conscious, and are more sensitive to environmental and social concerns (Richards, 2000, 2013). More specifically, archaeology tourists demonstrate essentially the same characteristics as traditional heritage tourists, whose main motives are to learn or to experience cultures and localities different from the ones where they live and spend time with loved ones (Alazaizeh et al., 2016; Blasco López et al., 2020; Etxeberria et al., 2012). In reality, the heritage and archaeology market is more complicated, with some destinations reporting younger demographics among international visitors, while domestic cultural tourists are more consistent with the traditional 'older' market profile (VisitBritain, 2010). Interestingly, the Canadian Tourism Commission (1999) identified four types of cultural and heritage tourists, namely cultural explorers, cultural history buffs, personal history explorers and authentic experiencers. These four types typically represent at least 30% of key international and domestic source markets and demonstrate different demographic and motivational characteristics.

Cultural and heritage tourism brings many economic, social and environmental benefits to destinations, and it is widely viewed as a tool for development. Economically, cultural and heritage tourism contributes to fiscal diversification, tax receipts, the protection of cultural and heritage 'economic' assets, as well as providing a catalyst for entreprenuers and small businesses to engage with tourism and tourist markets. In addition to the preservation of local traditions, customs and cultures, heritage tourism can contribute to a sense of local identity and pride, encourage place and community beautification, create valuable and frequently 'authentic' experiences, and serve as a catalyst for increased investment in supporting tourism infrastructure and services. Finally, heritage tourism draws attention to historic sites, archaeological digs and cultural attractions that are important for local residents as well as tourists, which can further help engender a culture of preservation and conservation. Beyond economics, tourism can contribute to maintaining culture and help build social capital and generate positive patterns of behavior. It is evident, therefore, that the benefits of cultural and heritage tourism are widespread with the onus on tourist destinations and the tourism industry to market, manage and develop tourism in a sustainable manner.

Marketing Archaeology-based Tourism

Marketing is not just about advertising and direct promotions. It is also about communicating, creating and delivering experiences that customers (i.e. tourists) will appreciate and relate to others. Creating satisfied customers who have pleasant memories of their visits is an important part of destination or attraction marketing. Recent marketing research in heritage contexts has suggested that this can be done by helping people appreciate heritage places better by improving accessibility, realizing the goals of sustainability and economic growth simultaneously, enhancing authenticity, utilizing suitable technology and appropriate interpretive media and providing connections and partnerships between localities in the form of trails and routes, to name just a few. The rest of the chapter examines many of these and other marketing mechanisms in the context of archaeological heritage.

Inherent in the wider trends of cultural and heritage tourism is the niche of archaeological tourism. 'Archaeotourism' is a growing phenomenon in both developed and developing countries (Díaz-Andreu, 2013; Pacifico & Vogel, 2012; Pinter, 2005; Walker & Carr, 2013). Archaeological resources for tourism include archaeological sites, material collections from archaeological field studies, data and related records, as well as the findings from archaeological investigations. More broadly, archaeological resources contribute to humankind's common heritage and aid our understanding and appreciation of the past. In addition to their cultural, spiritual and aesthetic values, archaeological resources contribute to tourism in many ways. For example, from a marketing perspective, archaeological resources and sites can help attract new markets to destinations, serve as the basis for differentiating destinations from one another by adding appeal and enhancing each locality's competitive advantage, and serving as a branding mechanism. It can also provide a catalyst for wider destination development and become the foundation for the authentication of the destination experience. Archaeotourism can also help celebrate the cultural richness of a region's history, and as already noted in the context of broader heritage tourism, it has the potential to bring economic growth to communities throughout the world (Babalola & Ajekigbe, 2007; Giraudo & Porter, 2010; Gould & Pyburn, 2017; Ramsey & Everitt, 2008; Rowan & Baram, 2004).

Timothy (2014) argues that archaeology-based heritage tourism plays an important role in justifying the public relevance of archaeological digs with the authenticity of archaeological sites contributing to a strong sense of place identity and an equally compelling sense of a high-quality, and more often than not, unique visit experience (Pujol, 2004). This latter point is especially significant, with archaeological sites being unique to a specific location and historical context. This combination of characteristics provides an ideal means of destination differentiation and diversification from traditional sun, sea and sand forms of tourism and even from other types of cultural tourism. To enhance these characteristics and archaeology-based tourism's potential, marketing is a key activity that heritage destinations and archaeology attractions need to consider.

Rather than focus on single destinations, Philippou and Staniforth (2003) outline the benefits to be derived from archaeological routes, with maritime archaeologists in some states of Australia extremely active in creating maritime heritage trails. Such trails are commonplace in the context of wine tourism, heritage tourism and whiskey tourism (most notably in Scotland) with their effective marketing efforts facilitating the development of tourism in often inaccessible and peripheral locations (Timothy & Boyd, 2015). The development of such trails can lead to local and regional economic development by linking supporting infrastructure and services, such as small hotels, restaurants and gift shops.

Although such measures help overcome issues of access, uppermost in the minds of those developing archaeological tourism is the sustainability of the attractions or sites in question and the extent to which they can be developed in the long-term economic, social and environmental interests of both the resident community and the tourists. Many archaeological attractions and sites are fragile and non-renewable, making it especially crucial to manage the impacts of visitors. Walker (2005), for example, discusses the difficult balancing act between tourism development and tourist accessibility with the need to conserve and protect archaeological resources in the long term. Common through the sustainable tourism literature is the need to engage the salient stakeholders as a means to ensure representation across the entire industry and to provide a collaborative bottom-up consensus as to what is right economically, socially and environmentally (Chirikure et al., 2008). This ideal is not always possible, but every effort should be made for public stewardship to be as representative as possible. Interestingly, and partially driven by research or educational interest, archaeological sites have a tendency to attract a number of participant volunteers.

Although a positive characteristic of archaeological tourism, such eagerness, however, does not always exhibit a strong understanding of the needs of the archaeological 'tourist'. Where participant volunteers do contribute strongly, though, is with the education and interpretation of archaeological sites and objects to visitors with them serving as educational and interpretation ambassadors. With strong educational backgrounds, they have a tendency to provide reliable and accurate information and an ability to deliver appropriate interpretation, especially when there are sensitivities with regard to particular spiritual or cultural values (Chirikure et al., 2008). McGregor and Schumaker (2006) add that such ambassadors also have an ability to engage visitors in alternative frames of thinking about culture and history. The appropriateness of some archaeological sites for purposes of tourism is a very real question with some, such as war graves or sites of human remains, arguably unsuited to visitation, with privacy and respect being higher priorities.

With authenticity and the authentic experience being acknowledged benefits of archaeological tourism, one recent trend is how virtual and augmented reality add to the visitor experience. Walker (2005) provides a useful synthesis of the presentation of archaeological remains to tourists via the use of such tools with the aim of enhancing the educational and visit experience. From three-dimensional (3D) documentation (everything from site surveys to epigraphy), 3D representation (from historic reconstruction to visualization) and 3D dissemination (from immersive networked worlds to *in situ* augmented reality), the digitalization of archaeological sites and exhibits is becoming commonplace (Bruno et al., 2010; Gillings, 1999; Goodrick & Earl, 2004; Han et al., 2018; Pujol, 2004; Rueda-Esteban, 2019). Pletinckx et al. (2000) introduce the heritage presentation program at Ename, Belgium, whereby different virtual reality approaches bring to life archaeological remains, standing monuments and elements of the historical landscape for visitors. For many archaeologists and those involved in archaeological tourism, digital tools and techniques offer new hope to the often painstakingly complex tasks of archaeology – surveying, historic research, conservation and education, with tourists very much benefitting from the technologies available.

Central to any form of archaeological tourism development is how it is marketed and branded. Ely (2013) identifies the need for archaeologists, communities and the tourism industry to unite to create effective marketing. Understanding the needs, wants and expectations of visitors is integral to any marketing program with the four types of cultural and heritage tourist identified earlier in the chapter advocated by the Canadian Tourism Commission (1999) a suitable platform to start. More recently, Sarker and Begum (2013) used a similar approach to develop niche marketing for cultural and archaeological sites in Bangladesh. Understanding visitor motives, whether local resident visitors or tourists, is critical with 'evocative' marketing, 'mood' marketing and 'emotional experience positioning' – all now common with places proferred as authentic by virtue of the emotions they evoke (Morgan & Pritchard, 1998; Prentice & Andersen, 2000).

Archaeological sites are prevalent throughout the world with many, Stonehenge in the UK being just one, representing iconic images frequently used in marketing and branding identities of tourist destinations to both domestic and international markets. The creation of positive images is an essential ingredient of destination market positioning, especially among close competitor destinations. In his study of Cyprus, Stritch (2006) outlines how archaeology is core to the Cypriot national identity and nationalist narrative, while it simultaneously contributes significantly to the promotion of the island as a leading destination. Archaeology also contributes to the diversification of Cyprus compared to other islands in the Mediterranean, as it also serves to differentiate among many destinations worldwide, especially those of an urban nature, with museums, galleries and exhibitions, and archaeological digs providing a unique, location-specific experience (Brida *et al.*, 2012). Many historic walled cities in England, such as Chester, York, Southampton and Winchester, are replete with archaeological remains that fulfill both an educative and economic role for local residents and tourists. In his study of Albania, Vladi (2014) also presents the benefits of archaeology to destination branding and image management, arguing that archaeological sites and exhibits may be a stimulus for destination loyalty.

One of the more evident roles archaeology plays in marketing is in the actual branding of destinations and in the shaping of destination identities. Kersel and Luke (2004) introduce the role of replicating archaeological sites in marketing tourism, and they shed light on archaeologists' and local government involvement in archaeological reproduction for promoting tourism. This features strongly in heritage branding by the International Council on Monuments and Sites (ICOMOS) and UNESCO for World Heritage Sites (see Ryan & Silvanto, 2010), just as it does in specific destinations. For example, Risitano (2006) outlines the destination branding process in the area of Campi Flegrei, Italy, a destination with a high intensity of cultural and landscape resources. There, a brand culture was created with archaeological sites at its core. Greece has long used its archaeological heritage for the purpose of destination branding as it promoted world-renowned sites with their associated social and political dimensions (Kavoura, 2012). This was evidenced at the Athens 2004

Olympic Games where organizers used Minoan archaeology to promote the event, as well as archaeology as theory, iconography and idealism to help celebrate its ancient traditions and modern organizational skills (Simandiraki, 2005).

Although not strictly an archaeological issue, the branding of Buddhism has caused controversy at the Bodh Gaya religious site in the Indian state of Bihar (Geary, 2008). Its designation as a World Heritage Site has resulted in considerable tourism growth, which not only brought to the fore the multiplicity of stakeholders often in competition within the destination, but in this particular case, the fear of spiritual degradation by tourism development projects. To address this shortcoming, it is critical to engage with the local resident community as the key stakeholder group.

In their study, Khirfan and Momani (2013) explain how the branding of Amman, Jordan, and its image influence, and are influenced by, the values Ammanis ascribe to their city. Addressing their values, images and identities of the archaeological tourism-focused destination was instrumental to the re-branding process and its ultimate success. Success is far from guaranteed, however, as Mortensen (2014) notes in his study of the branding of Copán, a Maya archaeological park in Honduras near the border with Guatemala. There, the case of '2012 phenomenon' was widely and mistakenly portrayed in the popular media as 'the end of the Maya calendar'. The issue of misrepresentation was also picked up by Magnoni *et al.* (2007) in their study of tourism in the Mundo Maya and how tourism can affect notions of self-identity and self-ascription (see also Medina, 2003).

The following example of the Parque Eco-Arqueologico in Huatulco, Mexico, illustrates the development of archaeological tourism, its rationale, benefits and challenges from the perspective of marketing and destination promotions.

The *Parque Eco-Arqueologico*: Archaeological Tourism in a Sun, Sea and Sand Destination

The state of Oaxaca is home to two of Mexico's 13 World Heritage Site-designated archaeological zones. Both are located near the state capital, Oaxaca City (itself a World Heritage Site): the ancient city of Monte Albán and the cave networks of Yagul and Mitla (UNESCO, 2017). While the cave networks are not open to tourists, Monte Albán is one of the ten most-visited archaeological sites in the country, averaging over 300,000–400,000 visitors annually (SECTUR, 2015). While the charms of Oaxaca City's colonial and historic buildings and streets are an important tourist draw, Monte Albán is considered a must-see day trip for all visitors to the region. Of the six additional archaeological zones developed for tourism near Oaxaca City (INAH, 2016), Mitla is one of the 20 most-visited archaeological sites in Mexico and a common stop on tourist excursions (SECTUR, 2015).

Oaxaca City is unquestionably the tourist center of the state, with nearly a million visitors staying at the city's hotels in 2014 (INEGI, 2015). The attractions of the city and surrounding area are almost entirely cultural, including archaeological sites, museums, music, traditional markets, cuisine and art. Thirty-four percent of the state's population speaks an indigenous language, primarily Zapotec (370,000 speakers) and Mixtec (260,000 speakers) (INEGI, n.d.). The traditional territories of both ethnolinguistic groups are centered near modern-day Oaxaca City, making for a simple narrative that connects the region's impressive ruins with modern descendent groups. Modern Native American groups' connections to Oaxaca's past are central to the state's tourism promotion, which often features Native art and costumes (Secretaria de Turismo, 2017).

Far less attention has been given to archaeological resources along the state's Pacific coastline. Only two coastal areas have been developed for tourism: Puerto Escondido and Huatulco. While Puerto Escondido has been an international destination since the middle of the 20th century, Huatulco was a greenfield development of the Mexican Fondo Nacional de Fomento al Turismo (National Fund for Tourism Promotion – FONATUR) beginning in 1984 (Brenner, 2005). Like FONATUR's four other destination resort developments, including Cancún, Huatulco was designed specifically to attract foreign tourists to luxury hotels (Brenner & Aguilar, 2002: 510), with only limited interest in incorporating cultural heritage resources into promotional efforts. Growing competition from other Caribbean destinations for North American SSS tourists, however, led FONATUR to begin investing more intensively in cultural resources both around its resort sites and elsewhere in Mexico starting in the 1990s (Brenner, 2005; Brenner & Aguilar, 2002). The best known of these promotional efforts is the Mayan Riviera along the Caribbean coast of Quintana Roo. Anchored by the resort city of Cancún, that region includes the archaeological parks of Tulum and Chichen Itza. Including archaeological zones in tourism development plans aimed to distinguish Mexican beach destinations from their Caribbean competitors, while also reducing infrastructure and economic strain on the increasingly urbanized coastal resorts (Brenner, 2005: 143).

The Parque Eco-Arqueológico Copalita

In Huatulco, FONATUR's investment in archaeological resources is limited to the Parque Eco-Archaeológico Copalita (Copalita). Opened for tourism in 2013, Copalita is a collaboration between Mexico's federal Instituto Nacional de Antropología e Historia (National Institute of Anthropology and History – INAH) and FONATUR (Copalita Museum, n.d.; SECTUR, n.d.). INAH manages all of Mexico's 187 developed archaeological sites and affiliated museums, including Copalita. INAH's (2015) mission is to preserve and disseminate archaeological, anthropological, historical and paleontological resources 'to strengthen the identity and memory' of Mexican society. It is fundamentally an educational and cultural entity that developed out of early 20th-century nationalist rhetoric (Heau Lambert, 2015: 1115). FONATUR (2016), on the other hand, is a mechanism for economic development that aims at sustainable growth and the modernization of Mexican tourist destinations. At Copalita, FONATUR provided funds for archaeological excavations, a site museum, walkways through the 81-hectare park, and the consolidation of pre-Columbian architecture (Copalita Museum, n.d.) (Figure 5.1). INAH provided personnel to direct archaeological excavations at the ruins and developed informational content. Now it manages the site (Matadamas Dias & Ramirez Barrera, 2010).

The result of an awkward alliance between agencies with two very different purposes, Copalita has effectively failed to forward the mission of either organization. Copalita received 10,266 visitors in 2014, of whom 35% were foreign (SECTUR, 2015). Total stays in Huatulco's hotels that year were 402,733, of whom 10% were foreign (INEGI, 2015). The site is therefore drawing in a higher proportion of foreign visitors than Mexican nationals. Some foreign visitors arrive at the site from cruise ships on day trips. Nonetheless, fewer than 3% of all hotel guests in Huatulco visited the ruins in 2014. While Copalita was the fifth most-visited archaeological site in Oaxaca in 2015, it received only a fraction of the visits received by

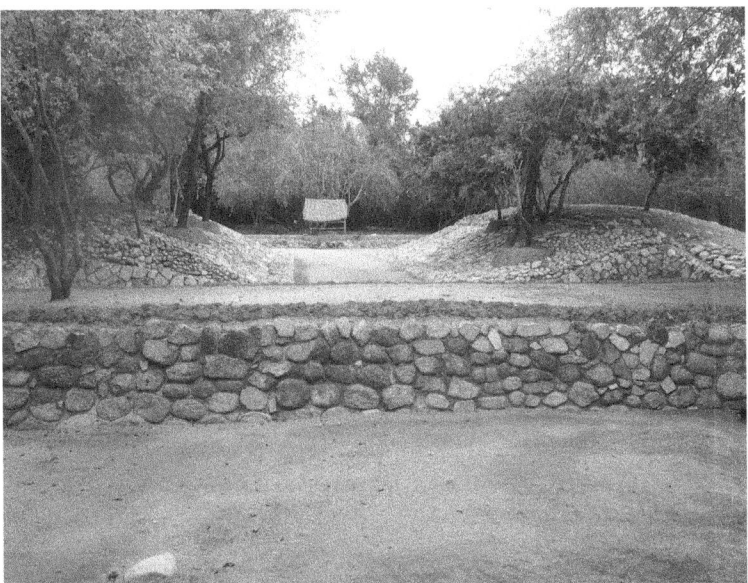

Figure 5.1 The Mesoamerican ballcourt at Copalita. These structures are popular architectural features among tourists and often a highlight of site tours in Mexico (Photo: Sarah Barber)

Monte Albán and Mitla and less than half the number of visitors at even smaller sites located near Oaxaca City. With little more than 30 visitors daily, Copalita is not drawing sufficient crowds to support many jobs. Many facilities (e.g. gift shops and cafés) remain unused.

As an educational facility, Copalita is even more unsuccessful. The park has a dual mission as a nature preserve. This means that very little of the forest that now covers the ruins has been cleared, making the minimally developed pre-Columbian architecture difficult to see. Sites elsewhere in Oaxaca are denuded such that archaeological materials stand out and ancient buildings' impressive sizes are not dwarfed by vegetation. Limited signage further reduces visitors' comprehension of what ruins are visible. The site museum is particularly problematic. While the building is beautifully designed, the content is largely disconnected from its geographic setting. Less than a third of the museum's display space is dedicated to the Copalita site. The remainder is divided into sections that focus on the archaeological materials of the Mixtec and Zapotec ethnolinguistic groups, with an emphasis on important artifacts from Monte Albán and other inland localities. Many of these pieces are reproductions, although they are not identified as such in museum signage. There is even a case displaying large stone knives that are likely Aztec – an ethnic group that had no pre-Columbian presence near Huatulco (Figure 5.2).

Archaeological sites near beach resorts elsewhere in Mexico receive enormous numbers of visitors, indicating strong national and foreign tourist interest in the country's ruins, even among beach tourists. The two sites located near Cancún are the second- and third-most visited archaeological sites in Mexico, with annual visitation surpassing 1.5 million at Tulum and 2 million at Chichen Itzá (SECTUR, 2015). Visits to Tulum equaled 24% of all hotel guests in Cancún in 2015 (INEGI, 2016). While the proportion of tourists in the region actually visiting Tulum is somewhat lower given the presence of hotels outside of Cancún proper, many more tourists there are visiting archaeological sites than are those in the Huatulco area.

Copalita's difficulties derive from both the promotional side and the educational side. The museum is less than five years old and likely needs more time to become incorporated into tour itineraries. It has also received very little promotion. For instance, there is no sign on the nearby highway indicating the exit to reach the site. Copalita also challenges established discourse around Oaxaca's pre-Columbian past. Archaeologists remain uncertain which modern indigenous groups are descendants of Copalita's builders (Pankonien, 2008). The most likely candidates are Zapotec speakers from the north or east, but speakers of other languages, including Chontal, Nahuatl, Chatino and Mixtec have historically occupied areas around modern-day Huatulco (Matadamas Dias & Ramirez Barrera, 2010; Pankonien, 2008). This ethnic diversity complicates narratives about the site's history, making it more difficult to connect the site to living peoples with whose traditions tourists might be familiar. Rather

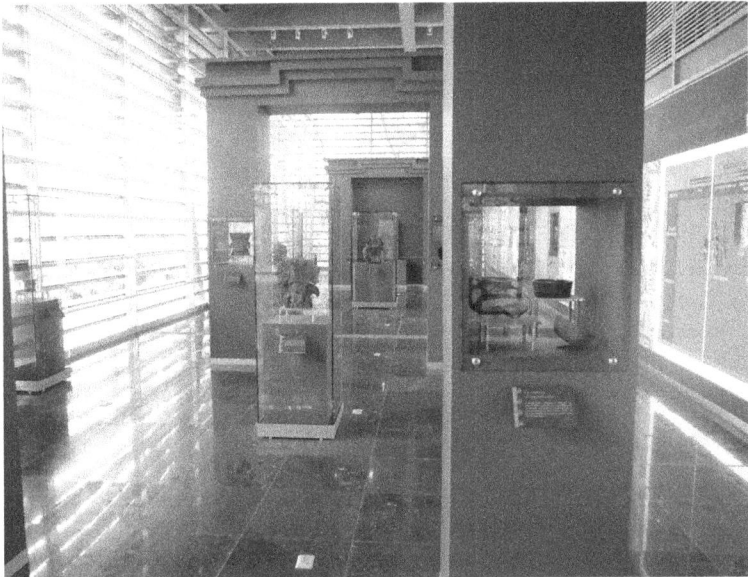

Figure 5.2 Interior of the site museum at the Parque Eco-Arqueologico Copalita containing reproductions of archaeological finds from elsewhere in Oaxaca state (Photo: Sarah Barber)

than embracing that ambiguity and linking Copalita to nearby, but less well-known, ethnolinguistic groups, the museum instead detaches the site from its surroundings and discusses the pre-Columbian history of highland Native American groups more familiar to tourists. The outcome is a hodgepodge of local, non-local, original and reproduced objects destined to disappoint most visitors. Domestic visitors will have seen the originals of many of the museum's reproduced pieces elsewhere in Mexico, while foreigners will not get the kind of legitimate local experience that lies at the heart of cultural tourism.

By way of a second example, one interesting and timely marketing campaign that has tapped successfully into the needs, wants and expectations of visitors is presented below as conducted by Historic Environment Scotland. As this case study demonstrates, an imaginative and creative marketing approach can facilitate the development of new markets and create positive engagement with history, heritage and archaeology, even among younger markets.

Historic Environment Scotland – the 'History Bug' Marketing Campaign

The lead public body set up to investigate, care for and promote Scotland's historic environment, Historic Environment Scotland (HES), has a variety of responsibilities that include conservation funding, education and serving as the custodian for 77 'paid for' attractions and over 300 unstaffed sites across Scotland. Many of these are archaeological sites of which many are in ruins. These include localities such as Skara Brae, Maes Howe and the Ring of Brodgar, which combine to form the UNESCO Heart of Neolithic Orkney World Heritage Site, as well as Scotland's busiest paid entry heritage attraction, Edinburgh Castle, centrally located in UNESCO's Old and New Towns of Edinbrugh World Heritage Site. Previously a government agency, HES became an executive non-departmental public body responsible for investigating, caring for and promoting Scotland's historic environment in 2016 when it was then renamed Historic Environment Scotland.

Historic Environment Scotland marketing

Marketing sits within both the Commercial and Tourism Directorate and the Business Development and Enterprise Department of HES. The focus is on marketing to visitors rather than other activities, such as conservation and funding, which sit within the wider organisation. Marketing activity is split into Consumer, Inbound and Travel Trade and Hospitality Events categories. Visitor numbers to the properties are split. Approximately 33% are residents of the UK, and 67% come from overseas, with European visitors contributing 28% of all visitors recorded. Working in partnership with the national visitor marketing body VisitScotland, HES participates in national campaigns such as themed years, #ScotSpirit, and promoting trails. HES's own campaigns in 2016 focused on the Year of Innovation, Architecture and Design, #HistoryHunters and #Bestdaysever. Focus is also given to driving memberships and events, many purely digital that lead to data capture for the Customer Relationship Management (CRM) program.

The majority of the marketing activity is directed to increasing membership sales and attendance at on-site events, while raising the profile of the sites through destination marketing organisation (DMO) partnership activity and a 'portfolio' marketing approach. This includes marketing clusters of sites in their locality in local area tourism guides, websites such as Welcometoscotland.com and visitor information centres. Membership and events tend to target the domestic Scottish market, as there is insufficient budget to market to the larger British market or internationally. As such, HES partners with VisitScotland and VisitBritain to promote the destination as a whole.

Particular care is taken in marketing the unstaffed archaeological sites. They are frequently not equipped to welcome large numbers of visitors, and protecting the fabric of the site from visitor damage is a priority. These archaelogical sites are frequently unstaffed, so there are no visitor control mechanisms in place. This can be problematic when visitor expectations are high, especially when visitors believe there is more to see than there actually is (e.g. Ring of Brodgar, Orkney). There is, thus, a dilemma with regard to encouraging visitation to fragile and unstaffed locations. That being said, sites such as Clava Cairns or Ruthven Barracks often prove popular and central to thematic marketing activities with Clava Cairns serving as a backdrop for Outlander, and Ruthven Barracks being integral to a project linking the Jacobite Exhibition at the National Museum of Scotland (NMS) with the Royal Collection. HES and the National Trust for Scotland encourage visits beyond the NMS exhibition through partnership marketing, predominantly digital via social media. Similar approaches are adopted by smaller regional sites, with thematic marketing providing a stronger message where there is less potential for so much timely messaging.

The 'History Bug' marketing campaign

The 'History Bug' marketing campaign was the winning proposal of an agency tender process held in December 2016. The brief was to create an imaginative sales-driven concept designed to capitalise on Scotland's Year of History, Heritage and Archaeology (one of VisitScotland's themed years). The basic premise was that 2017 was the best year to 'catch the history bug' and membership/events attendance is the only cure! The campaign was filmed in a 'family home' which kept production costs down and was documentary in style featuring a mother talking about how her son had 'caught the history bug' on a recent visit to Stirling Castle, only to realise as she speaks that she has it too! The online version of the campaign was slightly longer and featured a news anchor spoof who introduced the 'History Bug' epidemic (Figure 5.3).

HES worked closely with the winning agency to develop the concept through to campaign launch on 20 March 2017, the first time in seven years that HES had an offer to promote. Together, HES and the winning agency were able to negotiate 18 months of media space for the price of 12 months, so significantly more budget was invested into the use of television advertising to spread the message. This was then complemented by paid social media advertising and digital display advertising via a media buying agency, which used its knowledge of working with HES in 2016 to create an effective media plan. HES then used its own databases to promote the campaign along with its own organic social media content across eight channels. The membership element of the campaign lasted six weeks. Increasing membership sales to HES was the main objective, with the secondary objective of the marketing campaign being to enhance general awareness of HES attractions over the critical Easter holiday period. The entire tone of the campaign was informal and 'tongue in cheek' rather than serious. It was much more successful than was originally anticipated.

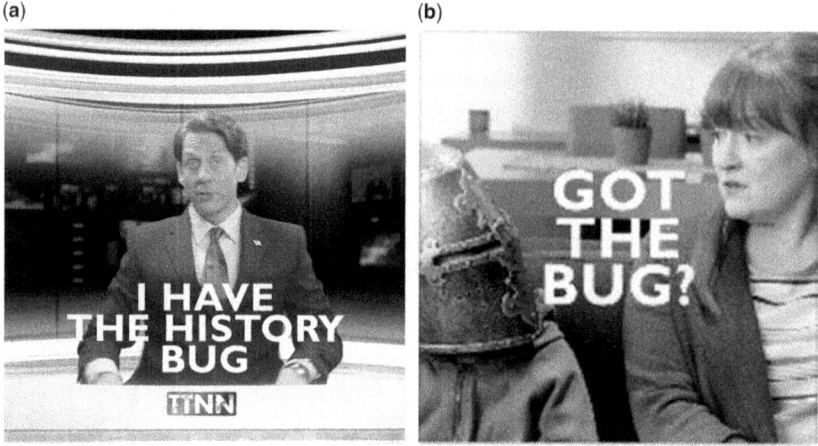

Figure 5.3 The 'History Bug' marketing campaign
Source: https://www.historicenvironment.scot/ (Images courtesy of Historic Environment Scotland)

It was also the recipient of a higher than usual budget, which facilitated access to a broader visitor market.

The marketing campaign closed on 30 April 2017, with HES delighted with the results. The desired membership sales target was reached at the end of week 5 with target membership numbers exceeded by approximately 37%. HES also recorded one of its most successful Easter visitation periods in recent years. While the dry weather and later Easter contributed in part, the raised profile of the HES brand through the 'History Bug' campaign also played a significant part. The television commercial was accompanied by a strong social media movement, with both paid and organic parts playing a prominent role in driving member sales and site visits. As with all such campaigns, the message was changed as appropriate. For example, once the primary Easter period had passed, an urgency was created that the offer was coming to an end. This was designed to maintain membership sales in the latter stages of the marketing campaign, which it duly did. Historic Environment Scotland planned to continue promoting the 'Bug' throughout the year as a mechanism to promote the events and membership, albeit this time without the offer.

Conclusion

This chapter sought to introduce the broader domain of archaeology and other elements of cultural heritage and the many benefits it can bring to tourist destinations before introducing the more specific contribution of tourism. As evidenced throughout the chapter, archaeological tourism is today part of a growing trend with strong growth, increased interest, and engagement from a variety of destinations and markets. The specific location and historical context of archaeological sites and attractions serve as a means to enhance the authentication of destinations with noted contributions to place identity and richness of the visit experience. It is, thus, no surprise that archaeological sites and attractions now contribute to the branding identity and imagery of many destinations, with their location and historical uniqueness serving as powerful tools for destination differentiation and competitive advantage. In turn, such sites and attractions serve to enhance the visitor experience with the preservation of local traditions, customs and cultures, local identity and pride, and sense of place, all contributing to an authentic visit.

The increasing use of virtual and augmented reality will do much to enhance the visit experience further, as well as provide a catalyst for attracting younger markets, similar to those attracted by the innovative and creative 'History Bug' marketing campaign in Scotland. Not only do these tools help bring archaeology alive, but they connect markets to their mobile worlds and enable positive electronic word-of-mouth and the sharing of travel experiences among peers on social media platforms. This is a critical component of modern tourism marketing.

In conclusion, it is evident that the benefits of archaeology-based heritage tourism are widespread with the onus on tourist destinations and the tourism industry to manage and develop this emerging tourism niche in as sustainable a manner as possible. Archaeological sites and attractions are often fragile and non-renewable and as such, it is imperative they are managed with care for the long-term economic, social and environmental interests of local resident communities and tourists. For this to be achieved, collaboration and sustainable marketing are essential (Chhabra, 2010) for effective and representative public stewardship and to guarantee the ability of archaeological destinations to meet the needs, wants and expectations of future generations of archaeological tourists.

Acknowledgements

Research funding for the Mexico case study was provided by the University of Central Florida.

Thanks to Lisa Robshaw, Marketing Manager Historic Environment Scotland, for making time to be interviewed and providing access to the 'History Bug' marketing campaign.

References

Alazaizeh, M.M., Hallo, J.C., Backman, S.J., Norman, W.C. and Vogel, M.A. (2016) Value orientations and heritage tourism management at Petra Archaeological Park, Jordan. *Tourism Management* 57, 149–158.

Babalola, A.B. and Ajekigbe, P.G. (2007) Poverty alleviation in Nigeria: Need for the development of archaeo-tourism. *Anatolia* 18 (2), 223–242.

Blasco López, M.F., Recuero Virto, N., Aldas Manzanob, J. and García-Madariaga, J. (2020) Archaeological tourism: Looking for visitor loyalty drivers. *Journal of Heritage Tourism* 15 (1), 60–75.

Brenner, L. (2005) State-planned tourism destinations: The case of Huatulco, Mexico. *Tourism Geographies* 7, 138–164.

Brenner, L. and Aguilar, A.J. (2002) Luxury tourism and regional economic development in Mexico. *The Professional Geographer* 54, 500–520.

Brida, J.G., Meleddu, M. and Pulina, M. (2012) Understanding urban tourism attractiveness: The case of the Archaeological Ötzi Museum in Bolzano. *Journal of Travel Research* 51 (6), 730–741.

Bruno, F., Bruno, S., De Sensi, G., Luchi, M.L., Mancuso, S. and Muzzupappa, M. (2010) From 3D reconstruction to virtual reality: A complete methodology for digital archaeological exhibition. *Journal of Cultural Heritage* 11 (1), 42–49.

Canadian Tourism Commission (1999) *Packaging the Potential: A Five-Year Business Strategy for Cultural and Heritage Tourism in Canada*. Ottawa: Canadian Tourism Commission.

Chhabra, D. (2010) *Sustainable Marketing of Cultural and Heritage Tourism*. London: Routledge.

Chirikure, S., Pwiti, G., Damm, C., Folorunso, C.A., Hughes, D.M., Phillips, C. and Pwiti, G. (2008) Community involvement in archaeology and cultural heritage management: An assessment from case studies in Southern Africa and elsewhere. *Current Anthropology* 49 (3), 467–485.

Copalita Museum (n.d.) Bocana del Río Copalita, Huatulco, Oaxaca. Museum signage. Viewed 11 May 2017.

Díaz-Andreu, M. (2013) Ethics and archaeological tourism in Latin America. *International Journal of Historical Archaeology* 17 (2), 225–244.

Ely, P.A. (2013) Selling Mexico: Marketing and tourism values. *Tourism Management Perspectives* 8, 80–89.

Etxeberria, A.I., Asensio, M., Vicent, N. and Cuenca, J.M. (2012) Mobile devices: A tool for tourism and learning at archaeological sites. *International Journal of Web Based Communities* 8 (1), 57–72.

Fondo Nacional de Fomento al Turismo (FONATUR) (2016) Misión, Visión. See http://www.fonatur.gob.mx/es/quienes_somos/index.asp?modsec=01-MV&sec=2 (accessed 19 May 2017).

Geary, D. (2008) Destination enlightenment: Branding Buddhism and spiritual tourism in Bodhgaya, Bihar. *Anthropology Today* 24 (3), 11–14.

Gillings, M. (1999) Engaging place: A framework for the integration and realisation of virtual-reality approaches in archaeology. *BAR International Series* 750, 247–254.

Giraudo, R.F. and Porter, B.W. (2010) Archaeotourism and the crux of development. *Anthropology News* 51 (8), 7–8.

Goodrick, G. and Earl, G. (2004) A manufactured past: Virtual reality in archaeology. *Internet Archaeology* 15, n.p. (online publication).

Gould, P.G. and Pyburn, K.A. (eds) (2017) *Collision or Collaboration: Archaeology Encounters Economic Development*. Cham, Switzerland: Springer.

Han, D.-I., tom Dieck, M.C. and Jung, T. (2018) User experience model for augmented reality applications in urban heritage tourism. *Journal of Heritage Tourism* 13 (1), 46–61.

Heau Lambert, C.M. (2015) Cuando la arqueología llega al recate del turismo: El caso de Bocana del Rio Copalita, Huatulco, Oaxaca, México. *Pasos: Revista de Turismo y Patrimonio Cultural* 13, 1109–1126.

Instituto Nacional de Antropología e Historia (INAH) (2015) About Us. See http://www.inah.gob.mx/en/about-us (accessed 19 May 2017).

Instituto Nacional de Antropología e Historia (INAH) (2016) Red de Zonas Arqueológicas del INAH See http://inah.gob.mx/es/2015-06-12-00-10-09/catalogo (accessed 17 May 2017).

Instituto Nacional de Estadística y Geografía (INEGI) (n.d.) Oaxaca: Diversidad. See http://cuentame.inegi.org.mx/monografias/informacion/oax/poblacion/diversidad.aspx?tema=me&e=20 (accessed 19 May 2017).

Instituto Nacional de Estadística y Geografía (INEGI) (2015) Anuario Estadístico y Geográfico de Oaxaca. Aguascalientes, Mexico: INEGI.

Instituto Nacional de Estadística y Geografía (INEGI) (2016) Anuario Estadístico y Geográfico del Estado de Quintana Roo. Aguascalientes, Mexico: INEGI.

Kavoura, A. (2012) Politics of heritage promotion: Branding the identity of the Greek state. *Tourism Culture & Communication* 12 (2), 69–83.

Kersel, M. and Luke, C. (2004) Selling a replicated past: Power and identity in marketing archaeological replicas. *Anthropology in Action* 11 (2–3), 32–43.

Khirfan, L. and Momani, B. (2013) (Re)branding Amman: A 'lived' city's values, image and identity. *Place Branding and Public Diplomacy* 9 (1), 49–65.

Magnoni, A., Ardren, T. and Hutson, S. (2007) Tourism in the Mundo Maya: Inventions and (mis)representations of Maya identities and heritage. *Archaeologies* 3 (3), 353–383.

Matadamas Dias, R.N. and Ramirez Barrera, S.L. (2010) *Huatulco: Antes de Ocho Venado y Después de los Piratas*. Oaxaca: Centro INAH-Oaxaca.

McGregor, J. and Schumaker, L. (2006) Heritage in southern Africa: Imagining and marketing public culture and history. *Journal of Southern African Studies* 32 (4), 649–665.

Medina, L.K. (2003) Commoditizing culture: Tourism and Maya identity. *Annals of Tourism Research* 30 (2), 353–368.

Morgan, N. and Pritchard, A. (1998) *Tourism, Promotion and Power: Creating Images, Creating Identities*. London: Wiley.

Mortensen, L. (2014) Branding Copán: Valuing cultural distinction in an archaeological tourism destination. *Journal of Tourism and Cultural Change* 12 (3), 237–252.

Pacifico, D. and Vogel, M. (2012) Archaeological sites, modern communities, and tourism. *Annals of Tourism Research* 39 (3), 1588–1611.

Pankonien, D. (2008) She sells sea shells: Women and mollusks in Hutulco, Oaxaca, Mexico. *Archaeological Papers of the American Anthropological Association* 18, 102–114.

PATA (2015) *The Connected Visitor Economy: The Role of Culture and Heritage Tourism in Building the Visitor Economy – And Beyond*. Visitor Economy Bulletin, April 2015.

Philippou, C. and Staniforth, M. (2003) Maritime heritage trails in Australia: An overview and critique of the interpretive programs. In J.D. Spirek and D.A. Scott-Ireton (eds) *Submerged Cultural Resource Management: Preserving and Interpreting Our Maritime Heritage* (pp. 135–149). New York: Springer.

Pinter, T.L. (2005) Heritage tourism and archaeology: Critical issues. *The SAA Archaeological Record* 5 (3), 9–11.

Pletinckx, D., Callebaut, D., Killebrew, A.E. and Silberman, N.A. (2000) Virtual-reality heritage presentation at Ename. *IEEE MultiMedia* 7 (2), 45–48.

Prentice, R. and Andersen, V. (2000) Evoking Ireland: Modeling tourism propensity. *Annals of Tourism Research* 27 (2), 490–516.

Pujol, L. (2004) Archaeology, museums and virtual reality. *Digithum* 6, 1–9.

Ramsey, D. and Everitt, J. (2008) If you dig it, they will come! Archaeology heritage sites and tourism development in Belize, Central America. *Tourism Management* 29 (5), 909–916.

Richards, G. (2000) Cultural tourism: Challenges for management and marketing. In W.C. Gartner and D.W. Lime (eds) *Trends in Outdoor Recreation, Leisure and Tourism* (pp. 187–196). Wallingford: CABI.

Richards, G. (2013) Culture and tourism: A naturally strengthening connection. Paper presented at the Board Failte, 2013 National Tourism Conference, Ireland.

Risitano, M. (2006) *The Role of Destination Branding in the Tourism Stakeholders System: The Campi Flegrei Case.* Naples: Department of Business Management, Faculty of Economics, University of Naples Federico II.

Rowan, Y. and Baram, U. (eds) (2004) *Marketing Heritage: Archaeology and the Consumption of the Past.* Walnut Creek, CA: AltaMira.

Rueda-Esteban, N.R. (2019) Technology as a tool to rebuild heritage sites: The second life of the Abbey of Cluny. *Journal of Heritage Tourism* 14 (2), 101–116.

Ryan, J. and Silvanto, S. (2010) World heritage sites: The purposes and politics of destination branding. *Journal of Travel & Tourism Marketing* 27 (5), 533–545.

Sarker, M.A.H. and Begum, S. (2013) Marketing strategies for tourism industry in Bangladesh: Emphasize on niche market strategy for attracting foreign tourists. *Researchers World* 4 (1), 103–107.

Secretaria de Turismo del Gobierno del Estado de Oaxaca (2017) Oaxaca: La Tiene Todo! See http://www.oaxaca.travel/index.php?option=com_content&view=category&layout=blog&id=10&Itemid=101 (accessed 19 May 2017).

Simandiraki, A. (2005) Minoan archaeology in the Athens 2004 Olympic Games. *European Journal of Archaeology* 8 (2), 157–181.

Stritch, D. (2006) Archaeological tourism as a signpost to national identity. In I. Russell (ed.) *Images, Representations and Heritage: Moving Beyond Modern Approaches to Archaeology* (pp. 43–60). New York: Springer.

Subsecretaria de Planeación y Politica Turística (SECTUR) (n.d.) *Parque Eco-Arqueológico/Eco-Archaeological Park Copalita.* Park brochure obtained on 11 May 2017. In possession of the author.

Subsecretaria de Planeación y Politica Turística (SECTUR) (2015) Compendio Estadístico de Turismo en México. See http://www.datatur.sectur.gob.mx/SitePages/CompendioEstadistico.aspx (accessed 18 May 2017).

Timothy, D.J. (2014) Contemporary cultural heritage and tourism: Development issues and emerging trends. *Public Archaeology* 13 (1–3), 30–47.

Timothy, D.J. and Boyd, S.W. (2015) *Tourism and Trails: Cultural, Ecological and Management Issues.* Bristol: Channel View Publications.

United Nations Educational, Scientific, and Cultural Organization (UNESCO) (2017) World Heritage List. See http://whc.unesco.org/en/list/ (accessed 18 May 2017).

VisitBritain (2010) *Culture and Heritage Topic Profile.* London, VisitBritain.

Vladi, E. (2014) Tourism development Strategies, SWOT analysis and improvement of Albania's image. *European Journal of Sustainable Development* 3 (1), 167–178.

Walker, C. (2005) Archaeological tourism: Looking for answers along Mexico's Maya Riviera. *Napa Bulletin* 23 (1), 60–76.

Walker, C. and Carr, N. (eds) (2013) *Tourism and Archaeology: Sustainable Meeting Grounds.* Walnut Creek, CA: Left Coast Press.

6 Archaeological Heritage and Volunteer Tourism

Dallen J. Timothy

Introduction

There is a growing trend in the tourism industry where certain consumers eschew the traditional experiences of mass tourism and self-indulgence. Increasing numbers of people are gravitating towards niche forms of tourism that simultaneously satisfy their personal needs while doing something positive for others. While people continue to embrace these niches, many of these special interest tourisms have begun to resemble the traditional mass activities and travels that have long dominated mainstream tourism. For example, ecotourism can legitimately now be seen as 'mass ecotourism' (Weaver, 2005), and heritage tourism has long been a crucial component of mass tourism. Nevertheless, there is a perception among certain niche travelers that they are more part of the solution than part of the problem. Volunteer tourism is one of the best manifestations of this trend. It typically entails people traveling to give back to society or to the earth, owing to various deep-seated desires to give of themselves or to work for the betterment of society or the environment, although it is far from being free of problems and controversy.

One form of volunteer tourism that has been largely overlooked by tourism scholars and heritage specialists is the crossover between archaeology, cultural heritage management and tourism. Tens of thousands of people travel each year on their own expense to become involved voluntarily in archaeological work or more general heritage management. Their motivations are manifold and the contexts in which they volunteer are extensive, yet we know very little about the phenomenon of heritage and archaeology-based volunteer tourism. This chapter provides an overview of this phenomenon that is frequently overlooked within familiar essays on volunteer tourism. Within the broader framework of volunteer tourism, this chapter examines general market characteristics and looks at the role of archaeology volunteer tourism as a form of public archaeology. It also highlights some geographical perspectives on the phenomenon and looks at the functions of the tourism industry with regard to this increasingly important manifestation of volunteer tourism.

Volunteer Tourism

Volunteer tourism is an 'alternative' to mass tourism traditions and reflects a growing dissatisfaction with conventional mass-produced travel experiences that are more leisure in their orientation (Mostafanezhad, 2016; Stoddart & Rogerson, 2004), although it almost always includes a recreational element (e.g. sightseeing, sunbathing on a beach or visiting museums) together with humanitarian or environmental service (Wearing, 2001; Wearing & McGehee, 2013a). Volunteer tourism enables participants to become more socially embedded in the destinations they visit, compared to many other forms of tourism, and to co-create deeper experiences that in many cases change their lives for the better. While people have traveled for charitable purposes (religious missions, disaster relief) for centuries, volunteer tourism in its present commercial form began in the 1950s with the emergence of large-scale service agencies and NGOs that sought to help the world's needy improve their living standards.

Today, volunteer tourism is vast and widespread and involves organized travel experiences where people give of their time, labor and expertise to help alleviate suffering, protect the environment, or work for social justice. Volunteer tourists generally cover their own costs (transportation, food, lodging) or participate in fundraising events to help pay for their travel expenses. Although there is a strong sense of altruism associated with this form of tourism, research suggests that the travelers themselves benefit personally just as much from these experiences, or even more so, than the causes they aim to help (Wearing, 2001).

In fact, there is a wide range of motives associated with people traveling for volunteer purposes. Brown (2005) suggests two main mindsets in the realm of volunteer tourism: 'volunteer-mindedness' and 'vacation-mindedness'. The focus of volunteer-minded tourists is the service work they travel to accomplish; they devote most of their away time to charitable activities in the destination. Vacation-minded travelers, on the other hand, participate more in leisure pursuits, with only a small portion of the holiday time being spent doing volunteer work, perhaps a few days or even a few hours. Vacation-minded volunteer tourists resemble the 'shallow volunteers' identified by Callanan and Thomas (2005, cited in Wearing & McGehee, 2013b), while volunteer-minded tourists best resemble Callanan and Thomas' 'deep volunteer tourists'. Shallow volunteers may be motivated more deeply by personal interests, while deep volunteers seek more altruistic outcomes from their experiences (Wearing & McGehee, 2013b). It is likely, however, that a mix of both types of motivations inspires the majority of volunteer tourists.

Many studies have revealed self-interest motives to include cultural immersion, making friends, bonding with family members, experiencing and learning something new, living in another country, expanding one's worldview, developing personal and marketable skills, developing personal networks, practicing another language, increasing one's faith, and providing job/resume experience (Ron & Timothy, 2019; Wearing & McGehee, 2013b). On the altruistic side, people desire to deepen their commitment to various causes, give back to society or the Earth, promote human well-being, and make a positive impact on the world. Certain studies have found that age may have a bearing on the level of altruism associated with volunteer tourism, with older people tending to display greater levels of self-sacrifice in their charitable travel experiences, although this depends on many variables.

From a supply perspective, common volunteer tourism activities include building and constructing homes, caring for children, conserving and protecting certain ecosystems, mending trails or fences, promoting human rights and social justice, providing medical service and health care, promoting literacy, sharing religious beliefs, teaching a foreign language, assisting in agricultural production, and providing aid and post-disaster relief (Grout, 2009; Wearing, 2001; Wearing & McGehee, 2013a).

Like many other 'alternative' forms of travel, much criticism has been leveled at volunteer tourism. It is now fashionable to participate in volunteer tourism, as its altruistic veneer seems to suggest that participants are selfless servants. Current criticism suggests that it is quickly becoming just another form of mass tourism that is packaged and promoted for large-scale consumption (Guttentag, 2015; Wearing *et al.*, 2016). As a result, volunteer tourism, like other forms of tourism, results in negative ecological, economic and social impacts, as well as neocolonialist relationships, wherein volunteer activities may benefit the destination to some degree but are equally harmful as they flaunt paternalistic and inequitable associations that sometimes disempower local communities and keep them in a constant state of dependency (Henry & Mostafanezhad, 2019; McLennan, 2014). According to Bandyopadhyay and Patil (2017: 644), volunteer tourists may see themselves as 'white saviors' who have come to rescue the brown locals who are incompetent to survive on their own until the next group of volunteers arrives. This can exacerbate the north–south divide even further and develop dispassionate views of outsiders among the people on the receiving end.

Regardless of these and other criticisms, volunteer tourism is still presented throughout the industry as a noble, yet prestigious, endeavor that entails self-sacrifice on the part of its participants. In addition to the common pro-earth and pro-poor activities outlined above, an area of service tourism we know relatively little about, because it lacks research, is heritage and archaeology-based volunteer tourism (Projects Abroad, 2019).

Heritage Work and Volunteer Archaeology

The cultural heritage industries have a long history of relying on volunteer workers. This has been especially true in the context of museums, historic homes and archaeological sites (Holmes, 2003; Stamer et al., 2008; Timothy & Boyd, 2003). Benson and Kaminski (2014) and Smith and Holmes (2009) differentiate between people who volunteer in the heritage tourism sector where they live (tourism volunteering), and volunteers who travel away from home to help staff historic sites, archaeological digs and restoration projects (volunteer tourism).

The majority of those who volunteer at heritage places in their home communities tend to be older, relatively affluent, well educated, and retired (Deery et al., 2011; Orr, 2006; Rhoden et al., 2009). For many people, volunteering in museums, tourism information centers, archives, heritage attractions, or archaeological sites is a means of continuing their hobbies into retirement and an enjoyable use of free time in retirement (Rhoden et al., 2009). Many studies have examined people's motivations for volunteering in the heritage sector. Findings regularly suggest a mix of both intrinsic/self-oriented purposes and extrinsic/selfless motives (e.g. Chen et al., 2019; Deery et al., 2011; Rhoden et al., 2009).

From the extrinsic or altruistic perspective, helping others appreciate and understand heritage, helping to protect the past, promoting peace and intercultural understanding, and giving visitors an opportunity to have an enjoyable day out are important motives (Deery et al., 2011; Rhoden et al., 2009). The results of one study (Chen et al., 2019) suggest that heritage volunteers, being largely attracted by the opportunity to interact with the heritage on display, were also motivated by opportunities to contribute their time, skills and knowledge for the advancement of the community.

From the perspective of self-interest, heritage volunteering provides intellectual stimulation and lifelong learning opportunities, and keeps participants healthier in mind and body and promotes overall wellbeing (Fredheim, 2018). Other reasons identified by Deery et al. (2011) include personal satisfaction, increasing self-esteem, meeting like-minded friends, and feeling needed, which are all very salient concerns among retirement-age volunteers. Heritage volunteering is frequently a reflection of one's hobbies and interests (Richardson & Almansa-Sánchez, 2015; Timothy & Boyd, 2003), and according to Holmes and Edwards (2008), for many people, volunteering in the heritage sector is simply part of an extended visit – a notion echoed by the work of Stebbins (1996, 2004), Orr (2006) and Chen et al. (2019). Because of the overwhelming role of self-interests, some scholars have described heritage attraction and museum volunteers as self-interested rather than pro-social (Lockstone-Binney et al., 2010). In the words of Lockstone-Binney and her colleagues (2010: 444), '…intrinsic motivators are of the greatest importance in attracting volunteers, but extrinsic factors may be equally important in retaining volunteers'.

Many heritage agencies, such as English Heritage, the US National Park Service, the National Trust for Historic Preservation, Heritage New Zealand, and Heritage Council (Ireland), rely increasingly on volunteers to realize their goals (Benson & Kaminski, 2014; Jameson, 2003). Nearly all heritage organizations understand that volunteer staff members allow them to accomplish things they would not normally be able to do (Benson & Kaminski, 2014; Kaminski *et al.*, 2011). Engaging volunteers has a wide range of benefits. First, they provide a huge cost savings for heritage institutions. While employing volunteers may not be entirely cost-free (e.g. investments in training), the fiscal savings of utilizing volunteer participants for various tasks is a huge cost savings for heritage establishments. Second, volunteers have the potential to enliven the institution with compassion and energy. They can provide new perspectives in interpretive programs or event planning, and they can provide insight into operations and management as members of the community. Third, volunteers also provide an important element of outreach between the institution and the community in what Fredheim (2018) terms heritage 'democratization'. 'Volunteers not only improve museum quality, but also enrich the community through a variety of outreach programmes' (Stamer *et al.*, 2008: 203). This can help bridge the gap between site managers and the community by acting as advocates for both stakeholder groups (Stamer *et al.*, 2008). Finally, volunteers also enable curators, archaeologists and historians to devote their time to the heritage craft, while some unpaid staff attend to the hands-on portion of creating a positive 'customer experience' (Smithson *et al.*, 2018).

It is clear that heritage tourism volunteering is an extremely important means of managing, protecting, marketing and interpreting the past to heritage consumers. However, the main concern of this chapter is the phenomenon of people traveling to volunteer at archaeological sites throughout the world – archaeological volunteer tourism.

Archaeology-based Volunteer Tourism

Just like heritage managers in general, archaeologists realize the need to employ volunteers in undertaking their scientific fieldwork. This growing support by archaeologists, together with the increasing prominence of volunteerism in the tourism marketplace, has seen a rapid growth in archaeology-based volunteer tourism in recent years (du Cros, 2019; Möller, 2019; Timothy, 2011, 2014). Today, many archaeological endeavors require volunteer vacationers for a number of reasons. Foremost among these is declining budgets in recent decades, which means fewer public and private funds are available to undertake excavations. Archaeology is labor intensive, and labor has traditionally been the costliest part of the excavation process (Kaminksi *et al.*, 2011). Not only do volunteers save excavation labor costs, they usually pay a program fee, which can be an additional income source to help fund the research. Likewise, accepting volunteers can help archaeologists better meet their goals and meet their project deadlines with the additional staffing (Timothy, 2018).

Unlike many other forms of volunteer tourism, archaeology-based volunteer tourism faces a shortage on the supply side. Despite the growing interest in participating in volunteer archaeology, there is a limited number of excavations each year, and not all of them seek volunteers. This has led many people to travel abroad to participate in excavating opportunities that are unavailable in their home countries (Möller, 2019). Kaminksi *et al.* (2011) outline several reasons for this limited supply. First, there has to be archaeology available to explore. While our planet is covered in remains of the human past, not all of it is accessible or in locations that are currently open to archaeological work. Second, some countries are challenging to work in. Some of the most intense archaeological clusters are located in war-torn countries or in areas under rebel control, which clearly limits archaeologists' ability to work. As of 2020, several ancient Assyrian sites in Iraq were still not open to archaeologists. As well, some countries have extremely tight regulations or extremely loose regulations for digging and conservation, both extremes providing their own sets of challenges. Third, there must be adequate management and professional staff, money and skills to be able to conduct excavations. This is a common problem but seems to be prevalent in India, other countries in South Asia and many countries in Sub-Saharan Africa. Fourth, not all projects are willing to receive volunteer laborers. Finally, language barriers can prevent some people from volunteering, but by the same token, many people desire to volunteer in areas dominated by another language to make the experience more 'foreign' or 'cultural'.

Who participates?

Archaeological excavations require considerable physical exertion, although not all duties associated with a specific dig are necessarily hard. On-site activities include clearing dig sites, removing topsoil and vegetation, digging archaeological strata and unearthing artifacts, sifting, surveying and measuring elevations, cataloging and photographing, soil sampling, moving spoil, and backfilling once the work is finished (Benson & Kaminski, 2014; Kaminski *et al.*, 2011).

Not all archaeological work, however, requires time directly at dig sites. Some volunteer vacationers spend their holidays restoring historic structures, entire ancient villages or medieval urban neighborhoods (Benson & Kaminski, 2014; Grout, 2008, 2009; Hamed, 2017; Timothy, 2018; Wearing, 2001). Other indirect archaeology-based activities include safeguarding artifacts and resources by repairing fences and trails, helping to implement an interpretive plan, cleaning up a historic village, repairing architecture following a natural disaster, assisting a community in its efforts to glean socioeconomic benefits from archaeological tourism, or assisting in visitor management efforts.

Few empirical studies have been done to understand the demand for archaeology-based volunteer tourism. Thus, little is known about this specific market, but it is likely that the market for this form of heritage tourism is as varied as that for other heritage volunteers. However, there are a few unique caveats. While home-based heritage volunteers tend to be slightly older and retired, this demographic characteristic may be less prevalent in the context of archaeology, especially in experiences abroad. Excavation-based volunteers are likely to be younger, which can be explained in three ways. First, excavation work is strenuous, much of it requiring heavy lifting, squatting, kneeling and bending over. Some advertised opportunities include specific requirements about volunteers' physical abilities to carry out the necessary on-site duties. While this does not preclude all pensioners or seniors, it does create a propensity for younger people to volunteer.

The second factor to determine the market relates to formal education. Many dig participants are secondary school, college or university students, who volunteer as part of their field school or service learning requirements (Figure 6.1). Many volunteer vacation opportunities are still geared overwhelmingly toward providing service learning for archaeology, cultural resource management, geography and history students,

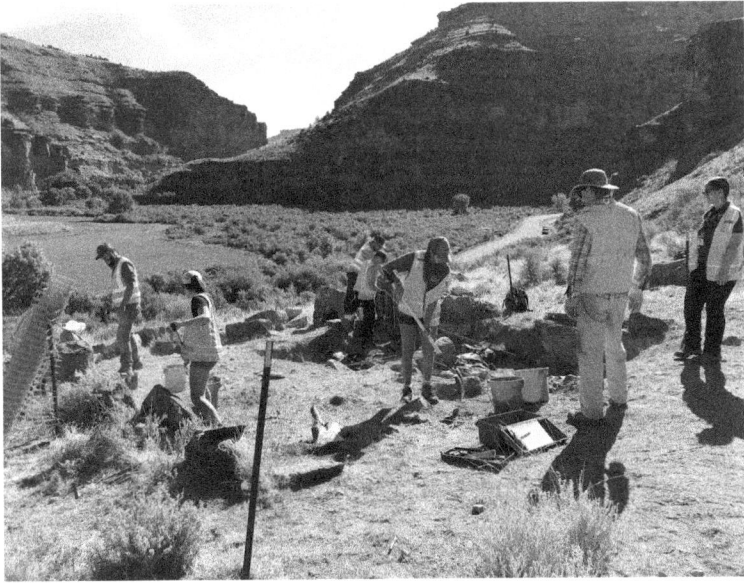

Figure 6.1 University and high school student 'tourists' volunteering on a Fremont Indian archaeological dig in southern Utah, USA (Photo: Dallen J. Timothy)

which combine digging and its associated activities with field trips, tours, and guest lectures (Geiger, 2004; Kaminski et al., 2011; Levine et al., 2005). Finally, past research has indicated that most adventuresome tourists, which includes archaeology volunteers, are a younger demographic (Pomfret & Bramwell, 2016), although retirees in the Western world are healthier, more affluent, and live longer than previous generations, which translates into their increased involvement in adventure types of travel.

While there are likely to be many self-oriented and altruistic motivations for participating in volunteer archaeology, the following sections examine three prominent motives: putting a hobby into practice, academic learning and religious devotion.

Pursing a hobby

People throughout the world have a wide range of personal hobbies and leisure activities that motivate them to visit certain places as tourists. For example, most countries are home to at least one postal history and stamp museum, which are significant attractions for philatelists, just as coin mints and engraving and printing agencies attract numismatists. Sport enthusiasts frequently visit historic stadiums and arenas, halls of fame and sports museums. People with interests in war history are drawn to participate in battlefield re-enactments, and historic railway enthusiasts not only participate in historic train trips but also avidly visit train and railroad museums (Timothy, 2011). Personal hobbies are, in fact, a significant incentive for personal travel, which is becoming increasingly common with the growth of specific niche forms of tourism (Novelli, 2005). Volunteering at home and during travel has also been conceptualized as a striking manifestation of one's leisure pursuits (Figure 6.2), which Stebbins (1996) has termed 'serious leisure'. Hobby-based niche tourism is arguably a form of personal heritage tourism and serious leisure, which also has implications for archaeology-based volunteer tourism.

As noted previously, heritage and archaeology volunteers often see their work as an extension of their hobbies. Volunteering for one or two weeks, or even an afternoon, can allow people to become involved in a passion for archaeology. Thus, while they might not have chosen archaeology as a profession, they can do it for a day or week as a hobby (Möller, 2019). One prominent example of a hobby area is militaria and military history. Military enthusiasts have traditionally had opportunities to participate in archaeological projects. One such effort in 2014–2015 allowed tourists to 'explore the Spanish Civil War' through heritage, archaeology and landscape by enrolling in the International Brigades Archaeology Project in Spain. The project aimed to understand better the approximately 30,000 warriors from 50 other countries who volunteered during the 1936–1939 Spanish Civil War to fight fascism and sustain the integrity of the Spanish Republic. These international volunteers became known as the International Brigades, many of whom gave their lives for the cause.

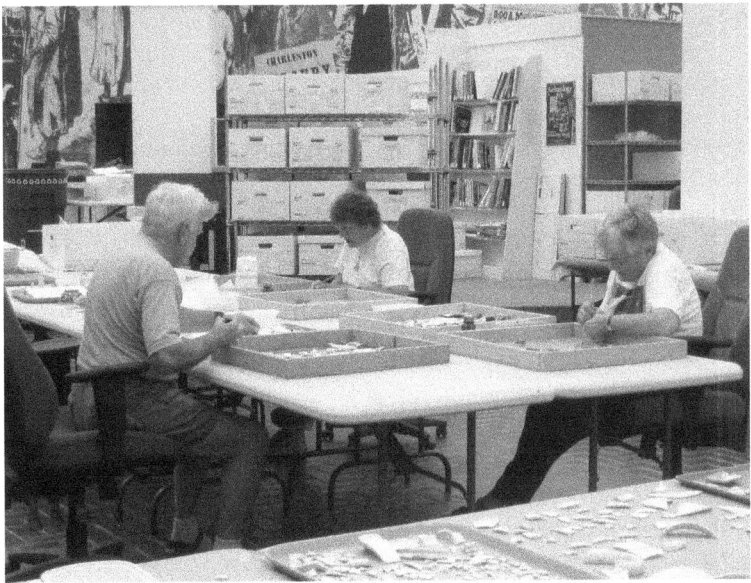

Figure 6.2 Archaeology enthusiasts volunteer their vacation time to help clean and catalogue artifacts at a museum in Philadelphia, USA (Photo: Dallen J. Timothy)

The International Brigades project was part of the larger Spanish Civil War Archaeology Project under the auspices of the Institute of Heritage Sciences of the Spanish National Research Council (Archaeology Fieldwork, 2019).

Another recent volunteer tourism opportunity for military enthusiasts was the Archeological Excavation at Woodford's Brigade site, a part of the larger Valley Forge National Historical Park Archeological Excavation in Pennsylvania, USA. This site is important in US history from the American Revolutionary War. The aim of the project was to define the limits of the Woodford Brigade encampment and to sample the remains of the site to provide a deeper understanding of everyday life during the encampment. The project was open to adults, children, teens and seniors, and entire families were encouraged to enroll to help screen for artifacts from excavated soil. On-the-job training was provided, with no prior experience or skills necessary (Archaeology Fieldwork, 2019).

Academic interests

Related to the hobbyist activities above is people's academic interests in certain time periods and events. While these events and periods might also be considered some people's hobbies, they might also be part of formal or informal educational pursuits (Stone & Molyneaux, 1994; Timothy & Boyd, 2003). From an informal educational perspective, volunteers might travel to participate in a Roman dig in Great Britain to enhance their own learning of British and/or Roman history. From a more formal educational perspective, university, college and high school students may travel to undertake volunteer archaeology fieldwork as part of their formal service learning curriculum.

During the past quarter century, a heightened interest in heritage truth-telling and democratizing heritage has resulted in more objective depictions of the history of slavery in the United States (Alderman *et al.*, 2016; Nelson, 2018; Timothy, 2011). This has, in part, resulted in extraordinary efforts to learn more about the enslaved through archaeology. One 2018 program invited volunteers to work with professional archaeologists at the Montpelier Mansion (Virginia) of James Madison, the fourth president of the United States and slave owner. The Excavating Slave Quarters at James Madison's Montpelier, Virginia project excavated domestic sites and work areas on the property of Montpelier Mansion, where the slaves of James and Dolley Madison lived and worked in the early 19th century. The goal of the program was to 'interpret and reconstruct the South Yard slave quarter so visitors can learn more about the African American heritage and contributions to the United States' (Archaeology Fieldwork, 2019: n.p.).

For volunteers interested in traveling to Belize to learn more about Maya culture and history, a 2019 opportunity allows them to focus on settlement patterns, rituals, ceremonies and water management among the ancient Maya. The focus of the Maya Commoner Archaeology in Belize project emphasizes geoarchaeological and environmental methods to understand soils and botanical remains to complement the work in previous years that focused on a ritual ballcourt and the rubble remains of a small house (Archaeology Fieldwork, 2019). Both of these opportunities are geared toward the formal educational pursuits of students, as well as the informal learning endeavor of non-student volunteers.

Religious interests

Biblical archaeology is a long-established subfield that investigates the material remains of the Bible lands (Holy Land and eastern Mediterranean) to shed light on events, times and descriptions in the Bible (Davis, 2004). Biblical archaeology has long attracted the attention of biblical scholars, theologians and religious adherents, many of whom become volunteer archaeology tourists for the cause of 'the gospel' (Etzrodt, 2012; Ron & Timothy, 2019). For example, there is a long history of Christians volunteering at religious heritage attractions in the Holy Land, such as the Garden Tomb and Nazareth Village (Ron & Timothy, 2019; Timothy & Ron, 2019). These volunteers, primarily from North America and Europe, work as guides, interpreters and actors, and are an important component of these sites' staff and make operating these sites a possibility.

From an archaeological perspective, there have been many Bible-related excavations throughout the Levant and eastern Mediterranean during the past couple of centuries. However, foreign tourist involvement began in earnest in the 1960s with the excavation of Masada, where hundreds of volunteer tourists, many of them Christians and Jews from around the world, were eager to become involved in digging in the Holy Land (Timothy & Ron, 2019; Yadin, 1966). Christians are avid volunteer archaeologists and are particularly keen on working at sites that support the biblical narrative. Many of them see these opportunities as spiritual experiences that draw them closer to deity and are manifestations of their faith in Jesus Christ (Ron & Timothy, 2019). Other recent and ongoing digs at Tell Azekah, Tell Keisan, Tell Dan, Tell es-Safi/Gath, Tell Shiloh, Bethsaida Excavation, Khirbat Safra, and Tell Hazor in Jordan, Palestine and Israel are all deeply connected to biblical events, peoples and places and are important dig sites for volunteer tourists (Cargill, 2019). Other Bible-connected sites in Egypt, Cyprus and Turkey continue to attract volunteer tourists from all over the world. The popular biblical Azekah dig is associated with the encounter of David and Goliath, which ostensibly took place nearby. In 2019, volunteers there required a two-week commitment and a fee of $500 per week excluding airfare. The project focused on understanding Late Bronze Age Azekah through digging, radiocarbon testing, residue analysis, petrography and ceramic analysis (Biblical Archaeology Society, 2019).

Archaeology-based volunteer tourism as public archaeology

Volunteer archaeology, whether at home or abroad, is a clear manifestation of the concept of public archaeology (Grout, 2009). Public archaeology, or community archaeology, utilizes community outreach and collaboration between archaeologists, site managers, community leaders and residents (Hoffman *et al*., 2002) to engage the public in archaeological research, making findings more accessible to the populace, and building public awareness of heritage through various archaeology-led efforts. In short, it has been said to be archaeology for the people, by the people. As archaeologists relinquish some degree of control over archaeology to the community, communities become empowered with a greater sense of solidarity and place-based social identity (Marshall, 2002) that will 'improve people's lives by helping them to enjoy and appreciate their cultural heritage' (Jameson, 2003: 161). This notion is predicated upon the principle that the public has a right to access its own past for learning purposes, identity formulation, and quality of life (Corbishley, 2011).

There are many examples of hands-on volunteering in Western societies serving to spark widespread public interest in archaeology (e.g. Grout, 2009; Jameson, 2003; Levine *et al*., 2005). This often entails the involvement of youth organizations, school or church groups, hobby clubs or families. In many cases, this refers to local activities revolving around local heritage. However, in the context of tourism, public archaeology also encompasses volunteer tourism, where the scale of 'community' is more global in its reach. Even regional outreach programs, however, can be seen as a mode of 'local tourism', formally so when groups stay overnight at the dig site and less formally when they undertake day-trips to a nearby excavation site.

Geographical perspectives

There is a unique geography associated with archaeology-based volunteer tourism. Thousands of people participate in volunteer vacations domestically each year, including on archaeological digs. This is particularly popular in the United States and the United Kingdom, as the higher number of archaeological volunteer opportunities in those countries indicates (Archaeological Institute of America, 2019; Archaeology Fieldwork, 2019; Möller & Karl, 2016). However, the lack of opportunities in some countries stimulates people from those countries to travel abroad to participate in archaeological opportunities (Möller, 2019), and the draw of being immersed in a foreign culture working in faraway places is part of the appeal of going abroad. Volunteer vacations are an essentially Western idea, in which few people from the developing world would be able to participate. General patterns indicate that most archaeology volunteers travel from developed countries to less-developed countries to participate (Hamed, 2017). Despite the overall dearth of active digs each year, the most plentiful opportunities are in Asia, the Middle East, the United Kingdom, the United States, Latin America and the eastern Mediterranean (Möller, 2019) (Table 6.1). There is a general lack of volunteer opportunities in sub-Saharan Africa owing to political problems and the prevalence of poverty rather than a lack of potential sites. Most African countries and Australia have few private societies to organize archaeological digs compared to other localities (Kaminski et al., 2011).

The Archaeology-based Volunteer Tourism Industry

With the growth of archaeology-based volunteerism, an entire tourism sub-industry has emerged with its own unique characteristics, suppliers and intermediaries. While the cultural remains themselves are the foundations of this phenomenon, scientific work would not exist without funders. In the past, much archaeology was undertaken through the financial support of wealthy individuals and/or government agencies. Today, however, private or non-profit organizations are the main sponsors of archaeological work. There are three main sources of excavation funding. The first are local societies funded through membership fees and grants. Second are developer-financed projects that have to be carried out before building permits will be issued. Third are educational agencies (e.g. colleges and universities) that fund digs with institutional budgets or grants for the

Table 6.1 Examples of archaeological excavations soliciting volunteer tourists for 2019

Project name	Country	Affiliation	Minimum stay	Main purpose	Main activities
Casa della Regina Carolina, Pompeii	Italy	Cornell University/ University of Reading	Five weeks	Excavation and survey of a large house in Pompeii	Excavation and identifying subsurface features by GPR
Town of Nebo Archaeological Project	Jordan	Wilfrid Laurier University	Six weeks	To examine ritual offerings in religion and landscape of ancient societies	Excavation and site survey
Paphos Theater Archaeological Project	Cyprus	Cyprus Department of Antiquities/ University of Sydney	Three weeks	Excavating a medieval structure to understand better the urban layout of the city	Excavation and cataloguing
Apollonia Pontica Excavation Project	Bulgaria	Balkan Heritage Foundation and others	Two weeks	Understand the Greek colonization of the Black Sea coast	Excavation and samples processing
Unearthing a Slave Community	USA	Montpelier Foundation	One week	Excavating an enslaved blacksmith complex and overseer's house site to understand the lives of the enslaved	Excavation and documentation
Western Mongolia Archaeology Project	Mongolia	National Museum of Mongolia/ Western Kentucky University	Three weeks	Investigate human–environment relationships and the social, political, and economic organization of Bronze and Iron Age societies in Mongolia	Excavation and use of geophysics and geoarchaeological methods
Manteño Structures in Agua Blanca	Ecuador	Universidad Técnica de Manabí	Two weeks	Understanding Manteño concept of the house and the social and symbolic significance of house architecture	Excavation and documentation
Stobi Excavations	Macedonia	Balkan Heritage Foundation and others	Four weeks	Excavate the most representative residential building in Stobi, the Theodosian Palace	Excavation and documentation
Rio Bravo Archaeology Survey	Belize	University of Texas at Austin/ Community College of Philadelphia	Two weeks	Investigate the remains of household structures, water systems, and ballcourts	Excavation, digital survey mapping, and laboratory activities

Source: Based on data in Archaeological Institute of America (2019).

purpose of scientific discovery (Kaminski *et al.*, 2011). These sources, through their generous financial support, inadvertently become part of the volunteer tourism sector.

There are hundreds of volunteer vacation sellers, but relatively few are devoted specifically to brokering archaeology experiences. Providers of the archaeology volunteer tourism product come in a few different forms. One is academic institutions, usually universities or research centers, who operate excavations and tender volunteer placements (Kaminski *et al.*, 2011). A second provider is the intermediary who packages the archaeology experience. Dozens of travel companies serve as clearinghouses for archaeology-based volunteer opportunities throughout the world. These firms collaborate with the funding agencies noted above, as well as governments and scientific research teams, to assess the human resource needs of specific digs and then proceed to market the opportunities worldwide (Timothy, 2018) (Table 6.2). The third common sort of provider are other types of associations, such as museums, trusts, anthropological societies, and foundations (Geiger, 2004; Kaminski *et al.*, 2011).

An example of a commercial intermediary is Archaeology Vacations. It is a private company that sells archaeological volunteer experiences, although its only client is the Lost City of the Manteños in Ecuador. According to the company's website, 'There are lots of tours around the world that will let you "look" at archaeological sites. We are the first company to actually let you excavate and discover the wonders of archaeology. This is not a university project looking for volunteers – Archaeology Vacations is specifically designed with the world traveller in mind' (Archaeology Vacations, 2019: n.p.).

Archaeology volunteer tourism is also less directly profitable for the destination than other forms of heritage tourism are. This is partly because it is an overall smaller market segment than most other types of

Table 6.2 A sample of companies that sell or promote archaeology-oriented volunteer vacations

Company name	Website (early 2019)
Adventures in Preservation	www.adventuresinpreservation.org
Archaeology Vacations	http://www.archaeologyvacations.com/
Balkan Heritage Field School	https://www.bhfieldschool.org/
Biblical Archaeology Society	http://digs.bib-arch.org/
Earthwatch Institute	https://earthwatch.org/expeditions
GoEco	https://www.goeco.org/area/volunteer-in-asia/cambodia/temple-preservation
Past Horizons	http://www.pasthorizons.com/worldprojects/
Projects Abroad	https://www.projects-abroad.org/volunteer-abroad/archaeology/

tourists. As well, general heritage tourists stay slightly longer in the destination than other tourists and are known to be high spenders (Timothy, 2011). However, volunteer heritage enthusiasts spend much less on average per day in the destination. Archaeology program fees are low compared to the prices of ordinary tour packages. Spartan lodging (e.g. on-site tents or bungalows, community centers, school gymnasia, or private homes) and most food are also included in the majority of archaeology programs, which means much of the archaeological volunteer economy remains outside the typical reach of government tax regimes and may provide relatively few local jobs. Thus, this type of tourism has much less fiscal impact than ordinary tourism, particularly as regards income acquired through accommodations and food services.

Conclusion

Archaeology-based volunteer tourism has received relatively little attention in the academic literature, compared to other types of volunteer tourism. While much empirical work is still needed to understand this phenomenon in greater depth, this chapter provides several insights from the perspectives of supply and demand, industry characteristics, geographical patterns, and its role as an important manifestation of public archaeology.

While we still know little about the market for archaeology-based volunteer tourism, some nascent patterns suggest that excavation participants may be younger overall, owing to the rigors of digging, the fact that service learning may be required for their formal studies, and that this activity is a form of adventure travel. Older volunteers may also work on digs, but may also be more inclined to help catalogue and clean, preserve, or work in an interpretive center. What we do know, is that one size does not fit all. Participation is determined by the specific activity or program, its location, its costs and its physical requirements.

In addition, it stands to reason that volunteer archaeology tourism may be less altruistic than other forms of volunteer tourism. Much of its focus is to satisfy self-interests, including hobbies, learning personally about places and periods, to develop faith and get more involved in one's religious heritage, to enhance a resume and one's employability, or to satisfy official curriculum requirements.

Another unique characteristic is that heritage and archaeological volunteer tourism does not appear to receive the same degree of criticism other forms of volunteer tourism receive. This is likely because this experience is less mass tourism oriented and characterized much less by the movement of do-gooders from the developed world to the developing world. Whereas most global volunteer tourists travel to less-developed regions ostensibly to help the poor and needy, the form of tourism described in this chapter exhibits higher levels of self-interest and typically requires less involvement with destination residents. Thus, by default it receives less condemnation for outcomes such as disempowerment, 'savior syndrome', and neocolonialism.

While archaeology-based volunteer tourism does not result in the same level of socioeconomic impacts other forms of tourism produce, it has proved its worth many times over. Some of the best-known archaeological discoveries were made possible through the work of volunteers. Hamed (2017) recognizes this in acknowledging that archaeology volunteers have helped save some of the greatest architectural wonders, ancient structures, monuments and old villages throughout the world. As word continues to spread about the value of volunteer archaeology and people's willingness to travel to participate, and if more opportunities become available, it will become more commonplace, and the volunteers will be in a better position to help protect the human past for future generations to come.

References

Alderman, D.H., Butler, D.L. and Hanna, S.P. (2016) Memory, slavery, and plantation museums: The River Road Project. *Journal of Heritage Tourism* 11 (3), 209–218.

Archaeological Institute of America (2019) Fieldwork. See https://www.archaeological.org/fieldwork/afob/search?field_afobtype_value%5B%5D=Volunteer&country=All&keys= (accessed 30 January 2019).

Archaeology Fieldwork (2019) Volunteer. See http://archaeologyfieldwork.com/AFW/Message/Index/0/6/1/volunteer (accessed 10 February 2019).

Archaeology Vacations (2019) Be an archaeologist on your vacation. See http://www.archaeologyvacations.com/ (accessed 12 February 2019).

Bandyopadhyay, R. and Patil, V. (2017) 'The white woman's burden': The racialized, gendered politics of volunteer tourism. *Tourism Geographies* 19 (4), 644–657.

Benson, A.M. and Kaminski, J. (2014) Volunteering and cultural heritage tourism: Home and away. In J. Kaminski, A.M. Benson and D. Arnold (eds) *Contemporary Issues in Cultural Heritage Tourism* (pp. 331–346). London: Routledge.

Biblical Archaeology Society (2019) Digs: Azekah. See https://www.biblicalarchaeology.org/dig/azekah/ (accessed 5 March 2019).

Brown, S. (2005) Travelling with a purpose: Understanding the motives and benefits of volunteer vacationers. *Current Issues in Tourism* 8 (6), 479–496.

Callanan, M. and Thomas, S. (2005) Volunteer tourism: Deconstructing volunteer activities within a dynamic environment. In M. Novelli (ed.) *Niche Tourism: Contemporary Issues, Trends and Cases* (pp. 183–200). Amsterdam: Elsevier.

Cargill, R.R. (2019) Digs 2019: A day in the life. *Biblical Archaeology Review* 45 (1), 1–9.

Chen, X., Liu, C. and Legget, J. (2019) Motivations of museum volunteers in New Zealand's cultural tourism industry. *Anatolia* 30 (1), 127–139.

Corbishley, M. (2011) *Pinning Down the Past: Archaeology, Heritage, and Education Today*. Woodbridge, UK: Boydell Press.

Davis, T.W. (2004) *Shifting Sands: The Rise and Fall of Biblical Archaeology*. Oxford: Oxford University Press.

Deery, M., Jago, L. and Mair, J. (2011) Volunteering for museums: The variation in motives across volunteer age groups. *Curator: The Museum Journal* 54 (3), 313–325.

du Cros, H. (2019) Planning and tourism. In S.L. López Varela (ed.) *The Encyclopedia of Archaeological Sciences*. Chichester: Wiley.

Etzrodt, C.L. (2012) Voluntourism: Motivations of the Bethsaida Excavation Project Voluntourism. Unpublished master's thesis, University of Nebraska.

Fredheim, L.H. (2018) Endangerment-driven heritage volunteering: Democratisation or 'changeless change'. *International Journal of Heritage Studies* 24 (6), 619–633.

Geiger, B.F. (2004) Teaching about history and science through archaeology service learning. *The Social Studies* 95 (4), 166–171.

Grout, P. (2008) *The 100 Best Worldwide Vacations to Enrich Your Life*. Washington, DC: National Geographic.

Grout, P. (2009) *The 100 Best Volunteer Vacations to Enrich Your Life*. Washington, DC: National Geographic.

Guttentag, D. (2015) Volunteer tourism: Insights from the past, concerns about the present and questions for the future. In T.V. Singh (ed.) *Challenges in Tourism Research* (pp. 112–118). Bristol: Channel View Publications.

Hamed, H.M. (2017) Voluntourism for preserving heritage: An initiative for safeguarding and developing New Gourna in Egypt. *Journal of Tourism and Hospitality Management* 5 (1), 34–45.

Henry, J. and Mostafanezhad, M. (2019) The geopolitics of volunteer tourism. In D.J. Timothy (ed.) *Handbook of Globalization and Tourism* (pp. 295–304). London: Edward Elgar.

Hoffman, T.L., Kwas, M.L. and Silverman, H. (2002) Heritage tourism and public archaeology. *The SAA Archaeological Record* 2, 30–33.

Holmes, K. (2003) Volunteers in the heritage sector: A neglected audience? *International Journal of Heritage Studies* 9 (4), 341–355.

Holmes, K. and Edwards, D. (2008) Volunteers as hosts and guests in museums. In K.D. Lyons and S. Wearing (eds) *Journeys of Discovery in Volunteer Tourism* (pp. 155–165). Wallingford: CABI.

Jameson, J.H. (2003) Purveyors of the past: Education and outreach as ethical imperatives in archaeology. In L.J. Zimmerman, K.D. Vitelli and J. Hallowell-Zimmer (eds) *Ethical Issues in Archaeology* (pp. 153–162). Walnut Creek, CA: AltaMira Press.

Kaminski, J., Arnold, D.B. and Benson, A.M. (2011) Volunteer archaeological tourism: An overview. In A.M. Benson (ed.) *Volunteer Tourism: Theoretical Frameworks and Practical Applications* (pp. 157–174). London: Routledge.

Levine, M.A., Britt, K.M. and Delle, J.A. (2005) Heritage tourism and community outreach: Public archaeology at the Thaddeus Stevens and Lydia Hamilton Smith Site in Lancaster, Pennsylvania, USA. *International Journal of Heritage Studies* 11 (5), 399–414.

Lockstone-Binney, L., Holmes, K., Smith, K. and Baum, T. (2010) Volunteers and volunteering in leisure: Social science perspectives. *Leisure Studies* 29 (4), 435–455.

Marshall, Y. (2002) What is community archaeology? *World Archaeology* 34 (2), 211–219.

McLennan, S. (2014) Medical voluntourism in Honduras: 'Helping' the poor? *Progress in Development Studies* 14 (2), 163–179.

Möller, K. (2019) Archaeologist for a week: Voluntourism in archaeology. In D.C. Comer and A. Willems (eds) *Feasible Management of Archaeological Heritage Sites Open to Tourism* (pp. 105–114). Cham, Switzerland: Springer.

Möller, K. and Karl, R. (2016) Digging up the past in Gwynnedd: Heritage research tourism in Wales. In G. Hooper (ed.) *Heritage and Tourism in Britain and Ireland* (pp. 229–243). London: Palgrave.

Mostafanezhad, M. (2016) *Volunteer Tourism: Popular Humanitarianism in Neoliberal Times*. London: Routledge.

Nelson, V. (2018) Object narratives and the enslaved at Sam Houston Memorial Museum. *Journal of Heritage Tourism* 13 (6), 235–249.

Novelli, M. (ed.) (2005) *Niche Tourism: Contemporary Issues, Trends and Cases*. Amsterdam: Elsevier.

Orr, N. (2006) Museum volunteering: Heritage as 'serious leisure'. *International Journal of Heritage Studies* 12 (2), 194–210.

Pomfret, G. and Bramwell, B. (2016) The characteristics and motivational decisions of outdoor adventure tourists: A review and analysis. *Current Issues in Tourism* 19 (14), 1447–1478.

Projects Abroad (2019) Discover what you're capable of. See https://www.projects-abroad.org/ (accessed 19 February 2019).

Rhoden, S., Ineson, E.M. and Ralston, R. (2009) Volunteer motivation in heritage railways: A study of the West Somerset Railway volunteers. *Journal of Heritage Tourism* 4 (1), 19–36.

Richardson, L.J. and Almansa-Sánchez, J. (2015) Do you even know what public archaeology is? Trends, theory, practice, ethics. *World Archaeology* 47 (2), 194–211.

Ron, A.S. and Timothy, D.J. (2019) *Contemporary Christian Travel: Pilgrimage, Practice and Place*. Bristol: Channel View Publications.

Smith, K. and Holmes, K. (2009) Researching volunteers in tourism: Going beyond. *Annals of Leisure Research* 12 (3–4), 403–420.

Smithson, C., Rowley, J. and Fullwood, R. (2018) Promoting volunteer engagement in the heritage sector. *Journal of Cultural Heritage Management and Sustainable Development* 8 (3), 362–371.

Stamer, D., Lerdall, K. and Guo, C. (2008) Managing heritage volunteers: An exploratory study of volunteer programmes in art museums worldwide. *Journal of Heritage Tourism* 3 (3), 203–214.

Stebbins, R.A. (1996) Volunteering: A serious leisure perspective. *Nonprofit and Voluntary Sector Quarterly* 25 (2), 211–224.

Stebbins, R.A. (2004) Introduction. In R.A Stebbins and M. Graham (eds) *Volunteering as Leisure/Leisure as Volunteering* (pp. 1–12). Wallingford: CABI.

Stoddart, H. and Rogerson, C.M. (2004) Volunteer tourism: The case of Habitat for Humanity South Africa. *GeoJournal* 60 (3), 311–318.

Stone, P.G. and Molyneaux, B.L. (eds) (1994) *The Presented Past: Heritage, Museums and Education*. London: Routledge.

Timothy, D.J. (2011) *Cultural Heritage and Tourism: An Introduction*. Bristol: Channel View Publications.

Timothy, D.J. (2014) Contemporary cultural heritage and tourism: Development issues and emerging trends. *Public Archaeology* 13 (3), 30–47.

Timothy, D.J. (2018) Producing and consuming heritage tourism: Recent trends. In S. Gmelch and A. Kaul (eds) *Tourists and Tourism: A Reader* (3rd edn) (pp. 167–178). Long Grove, Ill: Waveland Press.

Timothy, D.J. and Boyd, S.W. (2003) *Heritage Tourism*. London: Pearson.

Timothy, D.J. and Ron, A.S. (2019) Christian tourism in the Middle East: Holy Land and Mediterranean perspectives. In D.J. Timothy (ed.) *Routledge Handbook on Tourism in the Middle East and North Africa* (pp. 147–159). London: Routledge.

Wearing, S.L. (2001) *Volunteer Tourism: Experiences that Make a Difference*. Wallingford: CABI.

Wearing, S.L., Benson, A.M. and McGehee, N. (2016) Volunteer tourism and travel volunteers. In D.H. Smith, R.A. Stebbins and J. Grotz (eds) *The Palgrave Handbook of Volunteering, Civic Participation, and Non-profit Associations* (pp. 275–289). Basingstoke: Palgrave Macmillan.

Wearing, S.L. and McGehee, N.G. (2013a) *International Volunteer Tourism: Integrating Travellers and Communities*. Wallingford: CABI.

Wearing, S.L. and McGehee, N.G. (2013b) Volunteer tourism: A review. *Tourism Management* 38, 120–130.
Weaver, D.B. (2005) Mass and urban ecotourism: New manifestions of an old concept. *Tourism Recreation Research* 30 (1), 19–26.
Yadin, Y. (1966) *Masada: Harod's Fortress and the Zealots' Last Stand*. London: Widenfeld and Nicolson.

7 Archaeology and Religious Tourism: Sacred Sites, Rituals, Sharing the *Baraka* and Tourism Development

Nour Farra-Haddad

Introduction

The fields of religious sciences and archaeology are strongly connected, and the archaeology of religions and rituals is a growing research area. There is no doubt that understanding religions and rituals from past human societies helps us grasp the intangible religious past and developments in contemporary faith, beliefs and practices. Archaeologists study past cultures through material remains. The tangible heritage most commonly associated with the study of religion and spirituality include funerary objects and structures (cemeteries, tombs, mausoleums and necropolises), religious buildings (temples, churches, mosques, shrines and sacred caves), icons and objects of worship, and ritual tools from prehistoric times until today. Many such religious and funerary sites and structures have been inscribed on UNESCO's World Heritage List for their universal heritage value and serve as important tourist attractions throughout the world. Examples include the Egyptian Pyramids (eternal resting place of the Pharaohs), the temples of the Acropolis (Greece) and the cultural ceremonial center of Stonehenge (United Kingdom).

This book clearly demonstrates the close and historical correlation between archaeology and tourism. Also, scholarly research for decades has touched upon the relationships between tourism and religion, with increased attention being devoted to these relationships in the mainstream literature in recent years (e.g. Collins-Kreiner & Wall, 2015; Griffin & Raj, 2017; Raj & Griffin, 2015; Ron & Timothy, 2019; Stausberg, 2011; Timothy & Olsen, 2006). As a contribution to this growing research emphasis, this chapter focuses on the intersection between tourism, religion and archaeology. This tripartite connection transcends geographical and sociological emphases and involves economic (e.g. tourism) and political (e.g. national identities) dimensions. This chapter introduces the concept of religion and its intercourse with archaeology and tourism.

An analytical framework to understand the dynamics between religious heritage, devotional practices and religious tourism is defined. The chapter highlights the evolution of religious tourism and the importance of tangible and intangible religious heritage, and will focus these concepts on a handful of empirical cases from Christianity, Hinduism, Islam and Judaism with a special focus on the Middle East, particularly Lebanon.

This chapter invites readers to discover the importance of religious archaeological remains across the world and how they reflect the past. Tangible and intangible sacred heritage, especially archaeological heritage, offers amazing resources for tourism. Strategies to develop and promote religious tourism nowadays typically highlight the importance of the historical strata of religious sites.

The Context of Religion and Tourism

Religion is an age-old and dynamic concept; it is broad, abstract and complex. The study of religion encompasses many disciplines and engages a wide range of scholars, including archaeologists, sociologists, geographers, anthropologists, historians, philosophers, politicians and linguists, to name only a few. Postmodern discourses on religion cannot position its manifold concepts within one academic discipline and should 'question any possibility of rigid disciplinary boundaries' (Rosenau, 1992: 6). *Religionswissenschaft* (the science of religion) requires diverse disciplines to explore ideas, symbols and beliefs (Hinnells, 1984), as opposed to theology, which explores the truths or myths associated with religions, or with sociological discourses that insist on the importance of religion in creating social solidarity (Raj & Griffin, 2015).

The relationships between tourism and religion are many. Pilgrimage is usually considered the oldest form of non-economic travel (Jackowski & Smith, 1992). Every year millions of people of many faiths undertake pilgrimages to destinations around the world, both ancient and modern in origin. Religiously or spiritually motivated travel has become widespread, trendy and popular in recent decades, comprising an important segment of international tourism. The most traditional form of religiously motivated travel, the pilgrimage, is thought by some to be one of the forerunners of modern-day tourism (Cohen, 1992; Swatos & Tomasi, 2002). Travel for religious purposes is deeply rooted in the history of tourism.

Religious tourism is a niche sector within the larger frame of cultural or heritage tourism. It represents a very important component of the tourism industry nowadays and accounts for several hundred million faith-based journeys worldwide each year. In addition to its enormity, it is one of the fastest growing sectors within tourism (Hussain, 2016; Ron &

Timothy, 2019; Timothy & Olsen, 2006). Religious tourists include pious pilgrims with spiritual motivations, as well as secular visitors with cultural and curiosity motivations. Early on, Turner and Turner (1978: 20) acknowledged the intermixing of tourism and pilgrimage: 'if a pilgrim is half a tourist then a tourist is half a pilgrim', but as Nolan and Nolan (1992) and Smith (1992) explained, there is a wide range of visitor profiles with different interests and motivations within the broader religious tourism sector. Some tourists visiting religious sites are more interested in the historical significance of the shrine and its associated archaeological remains than they are in being moved spiritually.

Religious Stratification, Archaeology and the History of Sacred Sites

Over the centuries, the sacred was physically superimposed on the sacred, but new hallowed spaces also emerged. Churches were built over the ruins of earlier churches, pagan temples were converted to churches and some churches became mosques (Wallis & Blain, 2003). Some religious places have become popular attractions based on their archaeological remains. Thus, some ancient worship sites continue to be popular heritage attractions; many have been transformed, renovated or deteriorated, while other sacrosanct spaces have emerged more recently and quickly attracted believers.

Sacred geographies worldwide are evolving. Religious sites continue to be converted into shrines for other religions, and they continue to be culturally oriented. For example, San Lorenzo in Miranda occupies the remains of the Temple of Antoninus and Fostina in Rome; the Temple of Gaius and Lucius in Nimes was transformed into a church. Other former sacred sites have physically outlasted their sacrosanctity and are now considered simply archaeological sites visited only by curious tourists (e.g. the temples of Karnak, Delphi and Baalbek). From a different perspective, many new religious shrines have appeared in the modern era, enriching and expanding the world's sacred geography, such as the case of Our Lady of Lourdes (France), Our Lady of Fatima (Portugal) and Our Lady of Medjugorje (Bosnia and Herzegovina) (Carmichael et al., 1994; Ron & Timothy, 2019).

The phenomenon of pilgrimage is rooted in the past, since it commemorates religious figures, saints and important events; it is inseparable from history. However, pilgrimage is an evolving occurrence. Throughout history, many pilgrimage traditions have ended, for example during the Reformation in Europe, while others have emerged, as in the case of New Age travel. Just as pilgrimages evolve, over the centuries, sacred spaces have been built to overlap other sacred spaces, and new places have been sanctified. Some millennium cult places continue to be visited and transformed, converted and sometimes decline. Others will emerge and quickly attract the faithful. The development and evolution of religious sites is dependent on geopolitical and social forces, which can deeply affect or change, temporarily or permanently a place's sacred geography. There is ample evidence to suggest a life cycle process of discovery and the sanctification of places of worship, followed by their consecration and acceptance in the long term, and finally their transformation to profane spaces and their possible disappearance.

For example, in Lebanon, the site of Afqa 71 km northeast of Beirut is of interest to this topic. Afqa was an ancient pilgrimage destination. It has remained an important pilgrimage locality for the local Christian and Muslim population, and it has become an archaeological tourist attraction for foreign non-pilgrim tourists. The impressive entrance of the Afqa Cave, the 20-meter cave positioned in a 100-meter high cliff, and the spring of the Adonis River (Nahr Ibrahim River) once represented the power of the god Baal. This ancient landscape has inspired countless legends and traditions. In front of the cave, the remains of the Roman temple of Venus, shaken by earthquakes and eroded by time, testify of the importance of ancient pagan worship (Figure 7.1). In the foundation walls of the temple, a Marian shrine has survived with small houses, a small statuette of the Virgin where candles are lit, and textiles, clothes and rags are hooked on a century-old fig tree (Frazer, 1911, 1921).

This archaeological area of ancient pagan worship eventually became a venerated site for Our Lady of the Flowers (Saydet El Zahra) and Our Lady of Rafqa or Afka (Saydet Rafqa or Afqa), which is visited especially

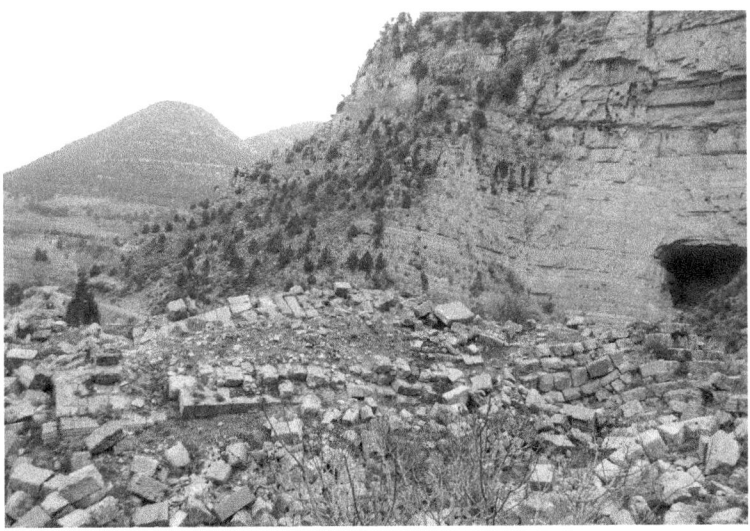

Figure 7.1 The Afqa Cave in rural Jbeil-Byblos, Lebanon, with the remains of the Roman Temple of Venus, shaken by earthquakes and eroded by time (Photo: Nour Farra-Haddad)

by women desiring to have children, which illustrates the modern Marian crossover with the ancient cult of Venus, the goddess of fertility. No religious authorities, Christian or Muslim, lay claim to the site today, but Lebanon's General Directorate of Antiquities is in charge of the site's protection and maintenance. The Directorate has enclosed the site within a fence and a locked gate, but these protective measures have failed to abolish the pilgrimage and rituals dedicated to the Virgin Mary. In spite of these barriers, the faithful continue to visit and breach the fence in order to carry out their pilgrimage activities.

In a very different context is the example of the church that was built in the 16th and 17th centuries on top of the ancient pyramid of Cholula in Mexico. The Great Pyramid of Cholula, also known as Tlachihualtepetl, is a huge complex. It is the largest archaeological site with a pyramid in the world. The pyramid is a temple believed to be dedicated to the god Quetzalcoatl and is closely linked to the ancient site of Teotihuacan. The Great Pyramid was an important pilgrimage center in pre-Hispanic times. Over a period of a thousand years prior to the Spanish Conquest, consecutive construction phases gradually built up the bulk of the pyramid until it became the largest in Mexico by volume.

During Spanish colonial times, the Church of Our Lady of Remedies (Iglesias de Nuestra Senora de los Remedios), also known as the Sanctuary of Our Lady of Remedies, was built on top of the pre-Hispanic temple, beginning in 1594. It is a major Catholic pilgrimage destination that is also used for celebrations of Indigenous rituals, in common with many such sites in Latin America. By the time the Spanish arrived, the pyramid was overgrown, and by the 19th century, it was still undisturbed, with only the church being visible. Because of the historic and religious significance of the church, the pyramid as a whole remains unexcavated and unrestored, as have the smaller but better-known pyramids at Teotihuacan. Inside the pyramid are some eight kilometers of tunnels excavated by archaeologists from 1881 revealing an intricate stratigraphy of different temple layers.

Archaeological excavations and historical studies can be conducted by religious communities to define the importance, long history, and holiness of a site, as for example was the case of the Orthodox Church of Saint George in downtown Beirut (Farra-Haddad, 2015a). After the Lebanese War (1975–1990), the Orthodox Church decided to conduct an archaeological excavation in the middle of the main nave of the devastated church to gather evidence of its being the oldest church in Beirut, or at least one of the oldest churches in Beirut. Nowadays, the crypt museum offers an amazing archaeological stratigraphy extending from the Hellenistic period to the Ottoman period, passing by the Roman, Byzantine and Medieval periods. The excavations showed that the church was indeed one of the oldest in Beirut built in the Byzantine period and restored many times afterwards. Archaeological excavations were similarly conducted under the Church of Saydet El Bourj (Our Lady of the Tower) in Deir El Ahmar, Lebanon, a few years ago, as was the case under the Cathedral of Our lady of the Seas in Tyre.

The archaeology of Jerusalem in general is as exciting as it is complex. Modern exploration there began in the 1830s, yet investigations are far from over. The archaeological investigations between 1960 and 1973 in the Church of the Holy Sepulcher, which was made possible by agreements among the different religious communities that control the church, were carried out to reveal its architectural history from its origins to the Crusader period. The church, which is also called the Church of the Resurrection or Church of the *Anastasis*, is located in the Christian quarter of the Old City of Jerusalem. According to traditions dating back to at least the 4th century, the church contains the two holiest sites in Christianity: the site where Jesus of Nazareth was crucified at a place known as Calvary or Golgotha, and Jesus' empty tomb, in which he is said to have been buried and from which he resurrected. The National Geographic-sponsored renovation of the traditional tomb of Jesus in the Church of the Holy Sepulcher in Jerusalem has offered new evidence that confirms the traditional links between the site and the events in the life of Jesus. Mortar recovered during recent renovations was dated to as early as 345 AD, using a scientific process called optically stimulated luminescence. This helps support the traditional dating of the construction of the first Church of the Holy Sepulcher, to mark the tomb of Christ, during the reign of the Roman emperor Constantine.

One of the most famous archaeological sites in the world, Baalbek, Lebanon, testifies of a very interesting religious stratigraphy. As early as 9000 BP, it was a place of worship and became a cornerstone of ancient civilizations. As a significant holy place, Baalbek was a center for Canaanite, Phoenician, Mesopotamian, Hellenistic, Roman, Christian and Islamic worship as each group successfully introduced its own heritage to this sacred complex. Through this process, Baalbek grew into an important pilgrimage site in the ancient world. The gods worshipped there, the triad of Jupiter, Venus and Mercury was grafted onto the indigenous deities of Haddad, Atargatis and a young male god of fertility. When Christianity was declared the official religion of the Roman Empire in 313 AD, Byzantine Emperor Constantine officially closed the Baalbek temples. At the end of the 4th century, Emperor Theodosius tore down the altars of Jupiter's Great Court and built a basilica using the temple's stones and architectural elements. The remnants of the three apses of this basilica, originally oriented to the west, can still be seen in the upper part of the stairway of the Temple of Jupiter. After the Arab conquest in 636 AD, the temples were transformed into a fortress, or *qal'a*, a term still applied to the Acropolis today, and a mosque was built inside the site.

As these examples illustrate, many archaeological sites around the world visited by tourists have a religious historical stratigraphy. For the faithful visiting shrines, this is frequently seen as an added benefit to discover that the site has a history and a long spiritual lineage.

Shared *Baraka* and Timeless Shared Rituals

Throughout the world, from antiquity until today, the faithful have followed votive pilgrimage routes and practiced votive rituals looking for '*baraka*' – the divine, miraculous spiritual presence of God in people's lives, sometimes delivered through the vehicle of saints but also to individuals according to their closeness to God. This force is thought to help the faithful in their everyday lives or in exceptional situations. The transmission of *baraka* operates mainly at sacred religious sites. For Fartacek (2012), in the context of Syria, the transmission of *baraka* is at the heart of shared ritual actions performed in the context of vows.

Most of the shared religious sanctuaries in the Middle East have very long histories that date back to pre-Christian and pre-Islamic times, and many rituals are inherited from antiquity. As the French 19th-century thinker, Ernest Renan (1997) wrote 'The sacred will replace the sacred'. Holy places are thought to contain *baraka* even when transferred from one religion to another; *baraka* is imbued to specific localities and remains on site and is inherited and transmitted over generations and through successive faiths. Many faithful also believe that the power of *baraka* may also increase with time. Pilgrims who perform *ziyārāt* in order to receive the graces that *baraka* offers also leave traces of themselves that can be observed by other visitors. It is often physical contact that is used to absorb the divine power of *baraka*: the pilgrim touches the statues, the wall of the temple, church or *maqām*, and/or touches or kisses the saint's tomb. Most of these ritual actions serve to transmit this divine presence and power. Pilgrims from all religious communities in the Middle East consider *baraka* to have a positive effect. The faithful look to it asking for protection in their daily lives or for help in exceptionally difficult situations. People visit holy places to obtain *baraka*, to make vows or fulfil them. To maximize the chances of a wish being granted, *baraka* must be obtained. The motives for visiting certain sites and performing certain rituals are grounded in the idea of *baraka*, as the faithful try to be imprinted with the benediction and grace of the holy site and carry this blessedness home with them.

Each sanctuary, shrine or temple prescribes a series of prayer rituals for the faithful. Rites and rituals are classified in order of importance by pilgrims depending on their potential effectiveness. Several rituals may overlap, follow, and be organized to make up a single votive approach or the framework for the same pilgrimage. Often, a main rite will be associated with other ritual practices, which crystallize to create an atmosphere favorable to the fulfillment of the desire. During the course of pilgrimages, devotees participate in a whole series of rites, including touching, ritual kissing of a tomb or sacred object, water rituals, contemplation, physical movements and prayers (Farra-Haddad, 2016) as a way of showing devotion and also evoking the power of *baraka*.

Excavations of religious sites all over the world have discovered that devotional practices and rituals have changed relatively little over thousands of years. Archaeologists use science and different methodologies to study ancient cosmologies, religious beliefs, intangible practices, and religious heritage sites, such as the upper Paleolithic cave art in Europe and the 'shamanism hypothesis' for interpreting it. Also studied are the Aztec calendar, British megaliths, ancient Egyptian funerary rituals, the famous Neolithic site of Çatalhöyük in Turkey, and the temple organizations of the Canaanite and Phoenician cities in Lebanon and around the Mediterranean Sea (Fagan, 1998). Analyses of these excavations help us understand contemporary religions and devotional practices, since so much of this was inherited from ancient traditions and beliefs.

Since antiquity, water rituals have had a very important place in pilgrimages throughout the world. Water is the lifeblood of humankind. It heals, rejuvenates youth, ensures life, and regenerates because of its purifying properties. Water is used for many types of rituals, including external ablutions and drinking practices. Ablutions can be done at the holy place directly, or holy water can be brought home for use later during symbolic moments (e.g. water ablution over a woman's belly before sexual relations). Water rituals have been practiced for centuries, and millennia, as the Canaanite, Phoenician, Roman and Byzantine remnants at the site of Eshmun, Lebanon, demonstrate. In the ancient world, from the Incas to ancient India, archaeological excavations reveal the importance of water rituals for devotional practices. Fallon and Jaiswall (2012) illustrate the importance of water in Hindu cosmology and its importance in religious pilgrimage tourism in India today. Water rituals are important in all religions of the world, including in Japan's temples, at one of the most important pilgrimage Christian sites in France: Our Lady of Lourdes (Caulier, 1990), and at the Jordan River where many Christian pilgrims are baptized or re-baptized in the place thought to be the baptismal site of Jesus (Ron & Timothy, 2019). The remains of fountains, wells, washbasins and water tanks at religious archaeological sites testify of the importance of water rituals throughout history.

Offerings or *ex-votos* are presented at the moment wishes are made or when thanks are offered. Offerings seal a pact between the faithful and the divine, an open request for help with the offering of a payment or appeasement to deity. Some observers see this as a commercial transaction and advocate wishes from the heart without offerings or 'contracts' with deity. Offerings usually manifest in two major types: valuable offerings or symbolic ex-votos. Many ritual objects, such as amulets, ablutions basins, alters and lanterns, are found in archaeological museums all over the world testifying that ritual devotional practices nowadays are very similar and shared by the faithful from different religious communities in sacred places.

Using Religious Archaeological Remains as an Asset for Tourism Development

A quick search of the internet shows that among the most visited religious places in the world are Mecca, Jerusalem and several archaeological sites, such as the temples of Luxor and Delphi. For Mecca and Jerusalem, there is no doubt about the archaeological heritage being a center of tourists' attention, and they continue to attract millions of devotees each year as important modern-day pilgrimage destinations. Luxor, Delphi and similar sites can be classified as ancient religious archaeological sites where devotion and religious rites are no longer practiced.

As several chapters in this book have made abundantly clear, archaeological remains have an important economic role to play, namely that of a resource for tourism. Places that mix archaeology and religion are frequently top tourist attractions. Most travel itineraries include visits to cultural sites that are somehow connected to faith – cathedrals, churches, mosques, temples and shrines. Many religious sites have a dual function as living places of worship and as tourist attractions. While this dichotomy tends to lead to different experiences for different types of visitors, Bond (2015) and other authors (Collins-Kreiner, 2010; Di Giovine, 2011; Koren-Lawrence & Collins-Kreiner, 2019; Olsen, 2012; Timothy & Olsen, 2006) have called for a shift away from dichotomizing the dual role of religious sites in favor of a deeper analysis of the complexity of this situation.

Trying to understand the diversity of the religious tourist experience, Bond (2015) conceptualizes the encounter by placing in the heart of the visit an interest in history, historical sites and culture. Visitors have distinct and overlapping motivations, interests and expectations, but most of them seek more than a casual encounter with an interesting or historic attraction. Regarding the overlap between tourism and pilgrimage, Stausberg (2011) and Ron and Timothy (2019) demonstrate the growing modern-day commercialization of pilgrimage and a growing interest among tourists in tangible and intangible religious heritage sites. The interface between tourism and pilgrimage is not only the tourist services, such as accommodations and food services, but also an interest in cultural heritage.

> Religious heritage sites enable visitors to feel part of something bigger than themselves and allow them to feel connected to both their histories and to other people in a way that visiting other heritage sites is unable to do... People, irrespective of their attitudes toward faith, find comfort in ritual, history and ceremony of the familiar. (Bond, 2015: 127)

Whatever people's motivations are to visit religious sites, an interest in heritage and history is shared by all, including pious pilgrims, for the sacred structures, rituals and traditions are a salient part of their personal heritage. The UNESCO World Heritage List is a highly coveted brand and a valued tool for developing tourism (Poria *et al.*, 2011; Stausberg 2011). Many localities on the list of World Heritage Sites are religious in nature with archaeological remains. Beyond organized religion are many places of spiritual strength that are venerated by 'spiritualists' and New Age worshippers as 'power places'. Many of the most famous of these are also inscribed on UNESCO's World Heritage List, including Cusco, Machu Picchu, sites in the Yucatan, Stonehenge, Mount Fuji, Petra and the Taj Mahal. This List 'reflects the process in which cultural resources of the world are perceived to be a part of the universal human heritage; the power places are something like a religious heritage list' (Stausberg, 2011: 98–99). Many of these places are sites of archaeological importance. Guidebooks regularly promote these localities, many have become iconic images of the tourism industries of the countries where they are located, and increasing numbers of specialized tour operators are developing circuits to these and other holy places.

Archaeological heritage also plays an important role in creating, defining and upholding national and religious identities. For many nations, the achievements of their ancestors are a focus for national identity building and represent a source of pride. Tourism uses material relics from the past in many different ways. Utilizing archaeological remains can support theories, causes and feelings of national belonging. In the modern-day Holy Land tourism product, Lebanon is frequently left out of the equation. However, scholars of tourism, geographers and historians are able to argue that Lebanon is indeed part of the Holy Land (Ron & Timothy, 2019). Archaeology plays an important role in identifying Lebanon as part of the broader Holy Land, and religious tourism in Lebanon could grow drastically if the country were able to be promoted as a part of the Holy Land. There are over 96 references to Lebanon in the Bible, and Jesus Christ himself is said to have walked on its soil. Christian communities have been present in Lebanon since the apostolic period, and prior to his detention, Saint Paul made many visits to Lebanon, often traveling through the coastal city of Tyre. In May 1997, Pope John Paul II proclaimed Lebanon as a holy land (and part of the Holy Land) for its privileged place in the Bible, its martyrs and its sacred places. Following this statement, many social initiatives and the Ministry of Tourism attempted to place Lebanon on the international tourism map of the Holy Land and encourage pilgrimages. Examples of organizations that worked towards this end include The Association for the Development of Pilgrimages and Religious Tourism in Lebanon, Lebanon, Holy Land and In the Footsteps of Jesus Christ in Lebanon. The General Directorate of Antiquities organized several digs in southern Lebanon in the vicinities of Qana and Qleileh, in areas known to be sacred. Excavation findings were analyzed and ostensibly confirmed visits by Jesus Christ in Lebanon (Farra-Haddad, 2015b).

In the village of Qana (Cana), archaeological remains from different periods have been discovered, some dating back to prehistory. The site (known as the 'Site of the Statuary') remained largely unknown for years, partly due to difficult physical access (the site is encircled by rocks), and only a few inhabitants of the village were aware of its existence. The rock carvings found on the walls date back to the early centuries, and depict characters in devotion, with hands lifted to the sky or close to the chest (Figure 7.2). Until the 1990s, this site remained without any facilities or signs to find it. A ministerial note dated 25 November 1993, stipulates the touristic and religious importance of 'Cana El Jalil' in southern Lebanon. Following the declaration of Pope John Paul II during his visit to Lebanon in 1997, the Minister of Tourism, Nicolas Fattouche, launched a major project to develop this site. However, the project remains unfinished and will take many years to be completed.

Behind the Maqâm of Nabi Omran, known to be the father of the Virgin Mary in the Muslim tradition, a field reveals an archaeological site. A local tradition mentions the existence of a Christian convent (*deir*) prior to the construction of the maqâm. On 27 April 1996, as a result of Israeli bombings, the remains of a crypt were revealed, which led to archaeological excavations by the General Directorate of Antiquities that identified mosaic floors and Byzantine structures, indicating the existence of a church with an apse. Several historical travelers and orientalists (e.g. Renan, 1997: 692) mention the ancient remains of this place.

Archaeological sites, excavations and discoveries can play an important role in sustainable tourism development. When a tourist hears that

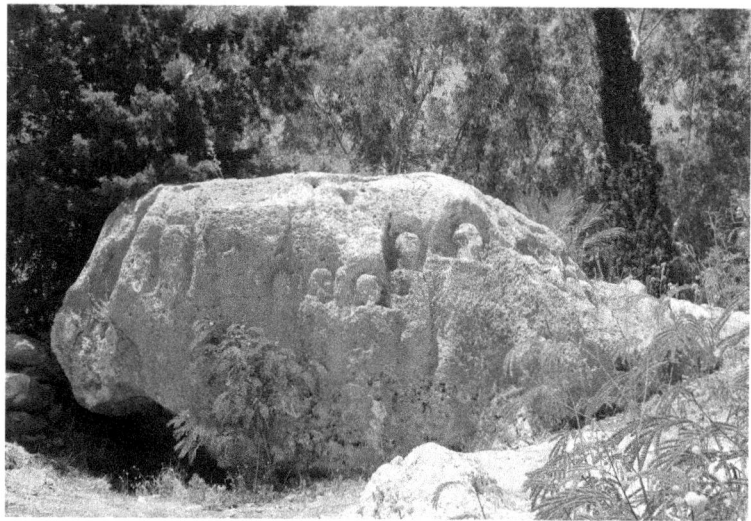

Figure 7.2 In the village of Cana, Lebanon, is the 'Site of the Statuary', where rock carvings date back many centuries (Photo: Noura Farra-Haddad)

the birthplace of the Buddha was discovered, or that the place of Jesus Christ's baptism was identified, he or she is exited to visit the site.

Walker and Tate-Libby (2012) demonstrate that the themes of archaeological and religious heritage are mediated and scripted by local communities to suit various agendas (economical, political and social). The articles in their special issue show how sacred sites become tourist attractions leading to control and management issues. Sacred sites that have been under the protection and management of local communities or religious authorities are usually reinterpreted, managed and controlled by outside tourism agencies or state organizations, creating environmental, economic, social and religious concerns. For instance, the level of sacredness of some localities might be conflated based on archaeological findings, which can create tension between scientific interpretations of an archaeological site and religious interpretations of the place (Walker & Tate-Libby, 2012). With similar concerns in mind, Koren-Lawrence and Collins-Kreiner (2019: 142–145) identified other areas of conflict associated with utilizing sacred archaeology for tourism. The first is the question of 'whose religion is it'. Dissonance between different groups that might lay claim to sacred places can create significant conflict. Second is the use of sacred archaeology as a political tool – to incite nationalism or uphold one group's claims or narratives over another. The third conflict results when scientific authenticity contradicts the 'symbolic authenticity' that religious tourists often seek and contradicts long-held local religious traditions. Fourth, contention may arise when the primary religious market for a specific site belongs to one faith, while the community that maintains and operates it adheres to a different faith not associated with the site itself. Finally is the instance where the aims of archaeology are at odds with the aims of pilgrimage and cultural tourism.

Regardless of these varied dissonances, sacred places will almost always be interpreted differently by secular tourists and by the faithful, but in all cases they are a source of attraction. Destination management organizations (DMOs) around the world use archaeological discoveries to promote their destinations and attract tourists to religious sites. Of the 845 listed World Heritage Sites, as of early 2019, most are archaeological remains and at the same time are connected to spirituality or religion and deemed sacred ancient places. Promoting archaeological heritage, particularly that of a sacred nature, is leading to an augmentation of visitors with different profiles to religious sites supporting tourism development. It also leads to an evolution in how these sites are managed and controlled. All these considerations raise many questions about the environmental, social and economic impacts of this development in sacred archaeological spaces.

Conclusion

Religious archaeology is a vast and dynamic subfield of research. Religious relics, remains and places are among the most visited heritage attractions in the world, and many have been inscribed on UNESCO's World Heritage List for their universal value and iconic representations. Such places exude a sense of awe and wonderment for ordinary tourists and devout pilgrims, but for many believers they are also intrinsically infused with a sacrosanctity (*baraka*) that heals, forgives, blesses and rewards in a way that transcends any particular religion and regardless of which faith the site currently represents.

While archaeological work has shown that many religious rites and rituals have remained essentially the same for hundreds of years, religious geography and archaeological sites are dynamic. Some undergo a particular life cycle of discovery, acceptance, veneration, decline and disappearance, in many cases only to be resurrected as a holy site by a different faith tradition later on. Many ancient places of pilgrimage and worship ceased their religious role over millennia or centuries but have become renewed spiritual destinations in Medieval or modern times, continuing to project their spirit of place. There are thousands of examples of newer or successive shrines, churches, synagogues and temples being built over the ruins of previous sanctuaries. Archaeology has been instrumental in demonstrating the steadfastness of faith traditions while simultaneously revealing the evolutionary nature of sacred space.

This chapter has focused largely on Christian archaeology because of its overwhelming emphasis and well-documented evidence in the religious and heritage tourism literature. Nonetheless, religious archaeology includes the remains and relics of many diverse religions throughout the world (Droogan, 2013). Many religious sites are the domain of faith organizations, archaeologists, devout pilgrims and other tourists. These diverse interests are known to create dissonance, even conflict, in owning, managing, protecting and visiting religious heritage places. Such is the case in Ayodhya, India, in which India's supreme court awarded a sacred plot of land to the Hindus in November 2019, on which a 16th-century mosque stood until it was destroyed by Hindus in 1992. The Hindus claim the site is the birthplace of Lord Ram and that local Muslims had built their mosque over the remains of an ancient Hindu temple nearly 500 years ago. The local Muslim population is challenging the court's decision; both groups contest this sacrosanct place and are attempting to use archaeological evidence to support their dissonant claims.

Archaeologists and their work are critical in uncovering vestiges of the past for visitors to experience and in providing an understanding of sacred geographies. Pilgrimages are a vital force in maintaining religious traditions while simultaneously accelerating change in the spiritual landscape. Despite some negative impacts, religious and cultural tourism at sacred sites can be useful in protecting the archaeological resources.

References

Bond, N. (2015) Exploring pilgrimage and religious heritage tourism experiences. In R.Raj and K. Griffin (eds) *Religious Tourism and Pilgrimage Management: An International Perspective* (2nd edn) (pp. 118–129). Wallingford: CABI.
Caulier, B. (1990) *L'eau et le sacré*. Paris: Beauchesne.
Carmichael, D.L., Hubert, J., Reeves, B. and Schanche, A. (1994) *Sacred Sites, Sacred Places*. London: Routledge.
Cohen, E. (1992) Pilgrimage and tourism: Convergence and divergence. In E.A. Morinis (ed.) *Sacred Journeys: The Anthropology of Pilgrimage* (pp. 47–61). Lanham, MD: Greenwood.
Collins-Kreiner, N. (2010) Researching pilgrimage: Continuity and transformations. *Annals of Tourism Research* 37, 440–456.
Collins-Kreiner, N. and Wall, G. (2015) Tourism and religion: Spiritual journeys and their consequences. In S.D. Brunn (ed.) *The Changing World Religion Map: Sacred Places, Identities, Practices and Politics* (pp. 689–707). New York: Springer.
Di Giovine, M. (2011) Pilgrimage: Communitas and contestation, unity and difference – an introduction. *Tourism* 59 (3), 247–259.
Droogan, J. (2013) *Religion, Material Culture and Archaeology*. London: Bloomsbury.
Fagan, B. (1998) *From Black Land to Fifth Sun: The Science of Sacred Sites*. Reading, MA: Helix Books/Addison-Wesley.
Fallon, J.M. and Jaiswal, N.K. (2012) Sacred space, sacred water: Exploring the role of sacred water in India's sacred places. *Recreation and Society in Africa, Asia and Latin America* 3 (1), 1–13.
Farra-Haddad, N. (2015a) Planning for sustainable tourism development in a context of regional instability: The case of the Lebanon. In N.D. Morpeth and H. Yan (eds) *Planning for Tourism: Towards a Sustainable Future* (pp. 186–202). Wallingford: CABI.
Farra-Haddad, N. (2015b) Pilgrimages toward South Lebanon: Holy places relocating Lebanon as a part of the Holy Land. In R. Raj and K. Griffin (eds) *Religious Tourism and Pilgrimage Management: An International Perspective* (2nd edn) (pp. 279–296). Wallingford: CABI.
Farra-Haddad, N. (2016) Shared rituals through ziyarat in Lebanon: A typology of Christian and Muslim practices. In I. Weinrich (ed.) *Performing Religion: Actors, contexts, and texts. Case Studies on Islam* (pp. 37–52). Beirut: Orient-Institut Beirut.
Fartacek, G. (2012) Rethinking ethnic boundaries: Rituals of pilgrimage and the construction of holy places in Syria. In G. Kilianova, C. Jahoda and M. Ferencova (eds) *Ritual, Conflict and Consensus: Case Studies from Asia and Europe* (pp. 119–130). Vienna: Austrian Academy of Sciences Press.
Frazer, J. (1911) *Le rameau d'or*. Paris: Schlincher frères.
Frazer, J. (1921) *Adonis*. Paris: Geuthner.
Griffin, K. and Raj, R. (2017) The importance of religious tourism and pilgrimage: Reflecting on definitions, motives and data. *International Journal of Religious Tourism and Pilgrimage* 5 (3), ii–ix.
Hinnells, J.H. (ed.) (1984) *The Penguin Dictionary of Religions*. Harmondsworth, UK: Penguin Books.
Hussain, H. (2016) *Islamic Tourism*. New York: Scitus Academics.
Jackowski, A. and Smith, V.L. (1992) Polish pilgrim-tourists. *Annals of Tourism Research* 19, 92–106.
Koren-Lawrence, N. and Collins-Kreiner, N. (2019) Visitors with their 'backs to the archaeology': Religious tourism and archaeology. *Journal of Heritage Tourism* 14 (2), 138–149.
Nolan, M. and Nolan, S. (1992) Religious sites as tourism attractions in Europe. *Annals of Tourism Research* 19, 1–17.

Olsen, D.H. (2012) Negotiating identity at religious sites: A management perspective. *Journal of Heritage Tourism* 7 (4), 359–366.
Poria, Y., Reichel, A. and Cohen, R. (2011) World Heritage Site: An effective brand for an archaeological site? *Journal of Heritage Tourism* 6 (3), 197–208.
Raj, R. and Griffin, K. (2015) *Religious Tourism and Pilgrimage Management: An International Perspective* (2nd edn). Wallingford: CABI.
Renan E. (1997) [1864] *Mission de Phénicie*. Paris : Impériale Printing.
Ron, A.S. and Timothy, D.J. (2019) *Contemporary Christian Travel: Pilgrimage, Practice and Place*. Bristol: Channel View Publications.
Rosenau, P.M. (1992) *Post Modernism and the Social Sciences: Insights, Inroads and Intrusions*. Princeton, NJ: Princeton University Press.
Smith, V. (1992) Introduction: The quest in guest. *Annals of Tourism Research* 19 (1), 1–17.
Stausberg, M. (2011) *Religion and Tourism: Crossroads, Destinations and Encounters*. London: Routledge.
Swatos, W.H. and Tomasi, L. (eds) (2002) *From Medieval Pilgrimage to Religious Tourism*. Westport, CT: Praeger.
Timothy, D.J. and Olsen, D.H. (eds) (2006) *Tourism, Religion and Spiritual Journeys*. London: Routledge.
Turner V. and Turner E. (1978) *Image and Pilgrimage in Christian Culture*. New York: Columbia University Press.
Walker C. and Tate-Libby J. (2012) The power of place: Heritage, archaeological and sacred. *Recreation and Society in Africa, Asia and Latin America* 3 (1), 1–3.
Wallis, R.J. and Blain, J. (2003) Sites, sacredness, and stories: Interactions of archaeology and contemporary Paganism. *Folklore* 114 (3), 307–321.

8 Archaeological Destruction and Tourism: Sites, Sights, Rituals and Narratives

Lina G. Tahan

> *Let us then, as much as possible, inscribe on*
> *all monuments and engrave in our hearts this maxim:*
> *'Barbarians and slaves hate science and destroy monuments*
> *of art. Free men love and conserve them'*
> (Abbé Henri Grégoire, 1793)

Introduction

It has long been acknowledged that archaeology is destroyed through different means. While some of it is unintentional, other forms are deliberate. Archaeological sites are constantly destroyed on a daily basis through human interference such as urban development, farming activities, highway and road construction, mass tourism or natural disasters such as earthquakes, tsunamis, and wind and water erosion. A major challenge for archaeological protection is that we cannot control what is being destroyed and how much of the archaeological record is lost. That we are unaware of what exists underground prevents local authorities from planning for what can be safeguarded for future generations.

Hence, destruction is inevitable, and while scientific excavations preserve a good record, the destruction is also irreversible. Archaeology is always a 'destructive' endeavour, but so is tourism through careless tourist behaviours, mass visitation, and infrastructure development (e.g. roads, car parks, gift shops) to provide services for the visiting public. This chapter discusses the destruction of archaeology and its effects on cultural tourism. It looks at man-made destruction from mass tourism, religious, terrorist, development and natural points of view. I argue that sometimes destruction is inevitable, affects the identity of the local community around a particular site, and influences the growth or decline of archaeo-tourism. In the latter case are archaeological sites that potential tourists might see as places damaged by vandalism and even violence and hence do not wish to visit. This chapter describes various forms of destruction of archaeological material culture and what this entails in the context of tourism.

Destruction through Mass Tourism

Tourism has changed the landscape of the world. There has been in recent times a dramatic increase of visitor numbers to famous archaeological sites. Without effective tourism management, it is not possible to preserve those sites for future generations. Tourism is a form of consumption, not of specific goods but of the experience of archaeological sites, and hence selling an experience has become a key part of the industry (Mitchell, 2002: 195). In many places around the world, the tourism industry has promoted the consumption of ancient remains. This can be seen in key locations such as Egypt, Greece and Mexico. Access to these places and the high number of tourists visiting them has caused tremendous problems for preservation efforts (Timothy, 2011).

At Petra, Jordan's Tourism and Antiquities Authority has implemented a very high entrance fee to ensure, in theory at least, that visitors at the site will appreciate the place and come out with a valuable 'experience'. The high fees help preserve the site. However, the local Bedouin community was displaced to develop tourism, which has caused clashes between the government and the Bedouins who have inhabited the caves for a long time, and their only means of sustenance was selling souvenirs to tourists (Comer, 2012). Relatedly, in Gurna, Egypt, the displacement of an entire community has caused discontent, and most of the blame was put on mass tourism and the fact that Egypt wanted to project a certain image of the site to fit foreign tourists' expectations (Meskell, 2005; Mitchell, 2002).

Another factor contributing to constant but non-deliberate destruction is abrasion, or wear and tear; the more people walk on stone, the more the masonry and the details on them are lost. Humidity fluctuations affect the interior of many enclosed archaeological sites. The grotto of Lascaux in France experienced severe damage to its wall paintings due to mould and green biofilm caused by the breath of large numbers of visitors. Hence, increased tourism led to the closure of the site in 1963 (Bastian & Alabouvette, 2009: 56). The solution in that case was to create a duplicate of the cave so that tourists could see and experience what prehistoric human paintings were like while preserving the original site after massive cleaning and conservation efforts. The facsimile, Lascaux 2, was only 200 meters away from the original grotto, but that put pressure on the original, and the French government decided to construct a new Lascaux 4 to ease the pressure on the original cave and its facsimile (Bryant, 2016).

Damage is also caused by vandalism and looting. Some tourists visit an archaeological site to loot artefacts or cause destruction on purpose. Hence, there is always tension between the preservation of archaeological monuments and mass tourism; this relationship has to be understood within the larger context of sustainable local development in order to prevent wear and tear on the archaeological sites. There is in effect a three-cornered triangle flanked by mass tourism, sustainable local development and the preservation of archaeological sites (Ashworth & Tunbridge, 2003). Each of these has its own individual problems and concerns, and when combined create problems for policymakers and heritage managers related to balancing archaeological heritage protection with mass cultural tourism to generate sustained economic development.

The Destruction of Archaeology through Religious Fanaticism

Archaeology is not only the study of the human remains of the past, but also a discipline imbued with politics, nationalism and legitimation of identity. Increasingly, it is used as a tool for propaganda and for achieving a specific agenda. The politics of governments are characterised by authority, power, legitimacy, control of resources and legality. All of these forces underlie a state's sense of national identity, and archaeology is frequently used to create a sense of belonging and a sense of power, and to shore up the national narrative. Once a site becomes a subject of conflict, there is reason to be concerned about its possible destruction. Such sites may become new hybrid places appropriated by various groups and subject to new interpretations. For example, Ayodhya in India became the location of religious violence and disputes in the 1850s over the Babri Mosque, because the Hindu community believed it was built on the site of the birth of the god Rama (Figure 8.1). The belief that the mosque was located on top of a temple spread among the Hindu community in the 19th and 20th centuries. In 1949, some Hindus brought Hindu idols inside the mosque and, as a result, the mosque was shut, no longer open as a place of worship. For the next 40 years, the situation remained tense until December 1992, when the mosque was completely destroyed. Because of this tragedy, both Hindu and Muslim communities lost their common Indian heritage and a sense of belonging. If the messages commonly hidden in nationalism are compared with, for example, ethnic or class groups, then it appears that the latter's concern is mainly with its distinct difference from other groups. They exist only in opposition to, or comparison with, those who are different. The emphasis is thus primarily on the present, and attention is devoted to creating a separate character, which can be experienced immediately here and now. It is not necessary that archaeology is involved in nationalism, but it is not surprising when it happens, and it is likely to spread through the use of religion. The example of India is one such case (Hole, 2013).

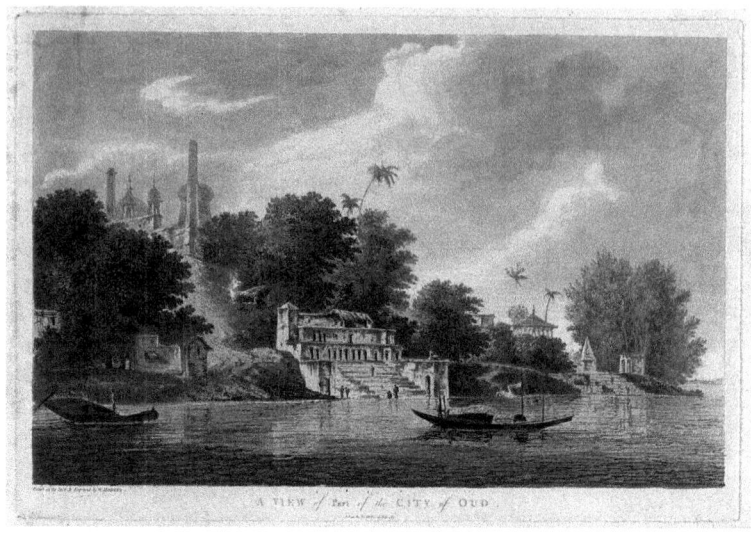

Figure 8.1 Ayodhya seen from the Ghaghara River, Uttar Pradesh, with the mosque depicted in the upper left. Coloured etching by William Hodges, 1785. Source: Wikimedia

When religious nationalism occurs, the distinction between one's group and the other is less central, and attention is rather focused on creating a bond between individuals and the nation. This means that efforts are made to make this relationship natural, to be taken for granted, and to give it emotional strength. Archaeology is in that context pliable; it can fit into this need. It is superbly suited to be used to provide a people's belonging and a sense of naturalness. It is, at the same time, a relationship that psychologically exploits the emotional impact of the sense of time, origin and ancestry and thus can have repercussions on the tourism industry. In that case, archaeological heritage is abused by destruction and used to fit or 'stand in' for the nation in various ways. This may happen in an abstract way, when the past is given the characteristics of the present state, both concretely in terms of geographic extent and more abstractly by embodying the character of the people/nation – industrial, communal and innovative. More specifically and importantly, archaeology becomes part of a wider narrative and becomes the means of transferring meanings and values of the past into the present, especially as regards the use of symbols. For example, the mosque in India, along with the temple, became a symbol of who had the power to dominate. Destroying it caused a certain malaise among the Hindu and Muslim communities, which has affected religious tourism to these sites and also resulted in retaliatory attacks in the city of Ayodhya. All of this could have been avoided had there been a government that had not played the Rama card and downplayed the secular ideals of the Indian State (Srivastava, 1994: 50). The same thing is happening with the Taj Mahal monument. Apparently, many in the current Indian government do not wish to list it as a main attraction in Indian guidebooks because it is part of the country's Muslim heritage (De Micheli, 2019). Such deliberate acts of alienating the heritage of the 'other' can lead to tensions at the national level. This leads into the following discussion about the abuse of the past in its various forms and why the so-called Caliphate of ISIS targeted old relics and archaeological sites that date back several millennia.

When important monuments are destroyed, the heritage of all of humanity is also destroyed. This can have detrimental effects on the tourism industry, especially if the monument is an iconic image of the destination, such as the Taj Mahal is for India. Such a situation tarnishes the destination's image and almost always results in a downturn in tourist arrivals.

Destroying Archaeology: Oppression, Greed, Vandalism and Carelessness, from the Taliban to ISIS

Recent conflicts in the Middle East and Afghanistan are closely related to identity struggles, and several agents have played a role in destroying museums and archaeological sites in that region (Ashworth & van der Aa, 2002; Butler, 2019; Meskell, 2015; Olsen, 2019). Destroying archaeological sites is a means of oppressing and controlling communities. ISIS, or Daesh, succeeded in imposing its effective oppression on communities living near important heritage sites. The impact on these people has been enormous, because this was a way for ISIS to impose itself and use famous tourist attractions to showcase its power and to demonstrate to the world that it cares nothing about the heritage of humankind. By destroying so-called 'pagan' sites, ISIS purported its form of Islam had supremacy and 'legitimacy' in the areas it controlled, despite the terror group's ideals being far from the more accepting and hospitable tenets of true Islam (Al-Kanany, in press).

The long list of archaeological sites destroyed while the world watched in horror dates back to 2001 when the Bamiyan Buddhas were blown up in Afghanistan (Ashworth & van der Aa, 2002). Then, in 2006, the Mosque of Al-Askari near Samarra, Iraq, was bombarded. In 2012, the old manuscripts and mausoleums of Timbuktu, Mali, were destroyed by the Ansar El-Dine and Al-Qaeda terrorist groups. Then came 2015 and the destruction of monuments in Syria and Iraq which became part of a daily ritual. The constant threat to cultural property became a part of socially-mediated terrorism (Al-Kanany, in press; Smith *et al.*, 2016). According to De Cesari (2015), the carefully staged destruction of Palmyra, Nimrud and Nineveh became the norms of terrorists' daily strategies of vandalism.

Smith *et al.* (2016) identified three strategies ISIS used to destroy ancient sites and relics. The first was to stage the destruction and test its impact. Second was to shock the international community, which felt helpless and unable to do anything. Thirdly, they looted artefacts to sell on the black market to finance their militant operations (Smith *et al.*, 2016: 164). Tourism played a crucial role in this endeavour. To place this in a larger context, beginning in the 5th century AD, fundamentalist Christians did something similar by destroying the icons of Pagan Egyptians with pickaxes and firebrands (Pollini, 2013: 241). Contemporary actions of fundamentalist Muslims are akin to similar actions by certain 5th-century Christians. ISIS adopted a ritual process of destroying pre-monotheistic cultural heritage (Shahab & Isakhan, 2018: 212). The use of Palmyra as a site of violence left a horrified international community condemning such barbaric acts. The 19 August 2015 staging of the torture and beheading of Dr Khaled al-Asaad, the former director of Syria's Department of Antiquities, showed the courage of an archaeologist who refused to reveal where many of Syria's archaeological treasures were hidden. ISIS' actions destroyed the idea of a common heritage and the sharing of knowledge, and imposed a single and linear version of history. Whatever preceded Islam, in the mind of ISIS, had no value and was to be destroyed because it went against the dogmas of this extreme form of Islam and its fight against idolatry and blasphemy. Hence, ISIS appropriated a past that it forged according to its own doctrine and the desire to reconstruct a Caliphate according to its political agenda (De Micheli, 2019; Jones, 2018).

As the above examples illustrate, in many cases, despite the significant roles they play in human development, identity and the nation state, monuments are non-renewable vestiges of the past that have long faced damage and deliberate destruction (Kirshenblatt-Gimblett, 1995; Layton & Thomas, 2009; Sørensen & Carman, 2009). While threats to Syrian and Iraqi heritage were real and ever-present (and remain so), nothing could be done by the international community or UNESCO. Apart from the loss of cultural property and the remnant materials of the past, archaeologists have observed that no matter what the motivation of ISIS and other terrorists was, we have failed to protect the past, which may lead to more harm as the development of culture in its present form has been stunted and marred by a loss of belongingness and identity (Lowenthal, 2015: 413).

Archaeologists are always thinking of the visiting public and how the archaeological record might be better brought into the domain of the public, including tourists. In the case of Syria and Iraq, ISIS destroyed famous touristic sites to show the world its will to annihilate ancient culture, tarnish the memories of those who have visited those famous archaeological sites, and prevent economic development through tourism.

The Destruction of Archaeology through Development Projects

As early as 1793 Abbé Grégoire wrote his *Report on the Destruction Brought about by Vandalism, and on the Means to Suppress it*, in which he denounced the destruction of cultural property. The destruction and damage barbarians caused in civilized areas have been called by several terms, including 'Degradation, Dissipation, Pillage, Mania, Destructive Furor, Mutilations, Frenzy Destruction, Assassination, Destructive Rage and Rascality' (Sax, 1990: 1161).

The words of this priest are echoed in the 21st century when destruction also occurs because of development projects and financial gains. Beirut, the capital of Lebanon, is one of the best examples that has endured archaeological destruction and irreversible damage since 1994. Once SOLIDERE (Societé Libanaise pour le Développement et la Reconstruction du Centre-ville de Beyrouth) took charge of the whole construction of downtown Beirut, things started turning sour between archaeologists and the company. Heritage and tourism professionals argued for the preservation of the ancient remains underneath modern buildings, suggesting that protecting and preserving them in situ would bring more value to the city. Developers were against preserving the city's archaeological remains because their financial interest clouded their thinking. The developers won the battle, and the ruins that were supposed to be preserved remain in ruins (Figure 8.2). There was no management plan to explain to residents or tourists what kinds of archaeology they could see. Instead, tourism's attention focused on the development of beautiful brand-name shops and cafés.

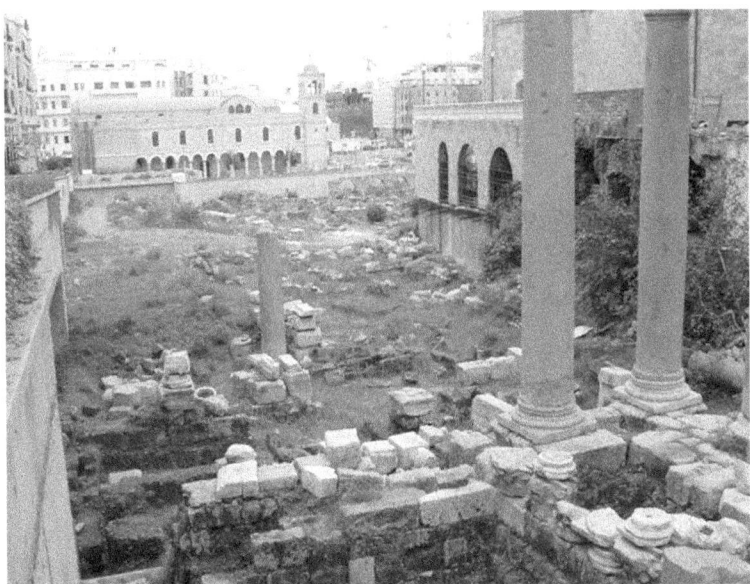

Figure 8.2 SOLIDERE wanted to create a 'garden of forgiveness' using the Roman basilica (Photo: Lina G. Tahan)

Such an oversight by developers left archaeologists puzzled as to why they could not use the past as an asset. One should bear in mind that it is very rare for a whole downtown area to have several digs open at one time. The Lebanese War (1975–1990) and its devastating effects helped pave the way to archaeological research. The government should have used this opportunity to protect and value its past, but unfortunately, things did not work according to plan. A few years later, downtown Beirut has become a virtual ghost city. Its beautiful buildings and high-rises are empty, because the Lebanese cannot afford to live there. 'SOLIDERE, despite its benefits, realised much of what people feared when the project began in 1994. Yet, despite its alteration of the city into a modern construction site and its erasure of the traces of Lebanese history, historical awareness has, in fact, taken a stronghold. SOLIDERE has become, furthermore, an overarching symbol for the destruction of the national history through construction' (van Pinxteren, 2018: n.p.) (Figure 8.3). This sums up how development can destroy archaeology, and there are thousands of other examples throughout the world of similar situations where the cause of 'development' has superseded the need to protect the past (McGill, 2003; Timothy, 2011). Although the website of SOLIDERE claims to have sponsored a heritage trail, this has not yet been implemented, and the company no longer has the cash flow that it once had in the early 2000s.

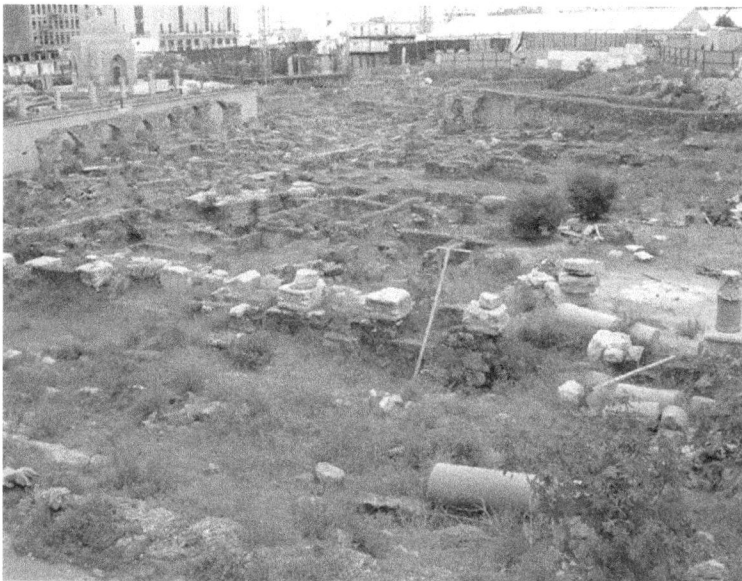

Figure 8.3 Contestation between developers and archaeologists has resulted in a stalemate in one of Beirut's archaeological areas (Photo: Lina G. Tahan)

Archaeology and Tourism: A Close Relationship

Archaeological sites are common throughout the world. Every country boasts an archaeological heritage as a matter of pride and as a mirror to the outer world in order to promote cultural tourism. A site thus becomes a place where the past is revisited and memory is recreated for a purpose. In his book, *The Heritage Crusade and the Spoils of History*, David Lowenthal (1998: 3) argues that representations of the past are a popular subject and that 'never before have so many been so engaged with so many different pasts' that nothing seems too recent or too trivial to commemorate.

This act of commemoration seems fundamental in the archaeological space and conveys several characteristics. One, it helps to recall public memory. Second, it triggers the visitor's memory to stimulate their thinking about important events in history and, consequently, introduces them to an unfamiliar or ignored personal or collective memory. That memory may be stimulated by looking at artefacts and engaging with historic spaces. This power of invoking memory is extremely important and is considered part of the cognitive aspects of human reactions in a specific cultural environment.

In a recent study, Crane (2000: 4) acknowledges that a museum is a 'repository of memory, location of collections that form the basis of cultural or national identity, of scientific knowledge and aesthetic value'. In this sense, museum exhibits, or for that matter, archaeological sites, become dynamic cultural intersections, where representations of the past closely and intimately interact with memory. Moreover, memory is not a passive process; it exists within the walls of museums – privileged places where the public can encounter what is being commemorated (Crane, 2000).

On these grounds, readers are reminded of the works of Halbwachs (1968, 1976), who pioneered the study of collective memory. This has been considered by many anthropologists, archaeologists, geographers and historians in their work, such as Nora (1984, 1989), Lowenthal (1985, 1998), Gathercole and Lowenthal (1990) and Bond and Gilliam (1994). When the archaeological record is destroyed, the collective memory of whole communities is likewise destroyed, along with the foundations of a country's tourism industry.

Conclusion: Annexing Culture

With its excavations of the distant and recent pasts, archaeology feeds the tourism industry with sites, sights and narratives. In some countries, archaeology has long provided a main resource for tourism development and a steady revenue source for governments. The relationships are frequently problematic between the tourism sector, with its emphasis on maximizing revenue from archaeological heritage sites, and archaeology as a practice of conservation and the scientific reconstruction of the past.

Archaeology serves the public good in various ways: research and discovery of the past, preserving collective heritage, interpreting and providing general education concerning history, artefacts and the habitats of ancient communities. Heritage tourism has become a buzzword in modern archaeology and in cultural studies. The debate concerning the concept of heritage and theories of its management has polarised much of the archaeological community and the tourism industry.

A discourse on archaeological heritage sites as a forum of interaction between professional communities and the general public has come into sharp focus over the past several decades. Archaeological excavations, historical ruins, battlefields, old buildings, walled cities, ancient roads and various cultural landscapes have become favoured tourist attractions, supported by the impact of movable museum collections and eventually the strength of narratives and image-engineering in a media sensitive society. The growing interest of a wide audience in heritage in general and historical sites in particular, explains the intensity of current debates on heritage and tourism, and the emergence of a new interdisciplinary field of scientific interest.

This chapter explored several forms of destruction and mentalities that are commonplace in contemporary archaeological and tourism discourses. Knowing that we live in a violent world full of wars and conflicts, destruction of cultural property seems inevitable. The archaeological record has long been, and continues to be, a target for militants who wish to inflict damage not only to the physical built environment but also to the national psyche and community solidarity that is often implicitly embedded within archaeological remains and in the stories they tell. It is important at this stage that heritage professionals act, raise awareness and ask questions, such as are you able to live without an archaeological site at your doorstop? A video was released recently by UNESCO and a campaign for #Unite4Heritage has released several videos about the destruction of human memory and the need for its protection for future generations (UNESCO, 2015). Hopefully, such efforts will raise awareness in heritage-hosting communities and among potential tourists about the importance of protecting the past for generations to come.

Recent theories on the politics and policies of cultural representations, cosmopolitanism, postmodernism and image conflict are addressed in various chapters of this book. The suggestion is that archaeology and heritage have much to offer the tourism industry by transporting archaeology beyond its traditions towards a dynamic expression of human understanding and thus becoming etched into the mental map of the tourist. The actual and potential transformation process of archaeological sites into heritage landmarks or icons and eventually tourist attractions requires full attention, not only from researchers in different disciplines, but also of regional and local authorities.

This chapter has examined the changing roles of stakeholders in this transformation process, with an emphasis on the development of integrated views and knowledge building. The dynamics of contemporary tourism development, policy and planning, implementation and management gradually enter into the scope of archaeologists.

Archaeological sites and museums are increasingly appreciated as cornerstones for the growing market of cultural tourism. The physical embedding of archaeological sites and museums into historic landscapes produces an aura of authenticity and adds value to the tourist experience. Tourism has become an important, if not the most important, supporter of the conservation of archaeological artefacts and sites. While maintaining a dedication to preservation, conservation and research interests of archaeologists, the intrusion of tourism activities, access and facilities, is an increasingly common practical and ethical concern. The objective to 'save the past for the future' is a valuable ethical concern and a real mission for many stakeholders; the optimal carrying capacity or limits on visits at particular sites need to be assessed realistically, by archaeologists, tourism agents and environmental experts.

Although tourism is known to be a destructive force against archaeological relics owing to masses of tourists and their careless behaviours, tourism also creates opportunities for archaeologists, visitors, local communities and tourism enterprises. However, the challenge is to define the dynamics of interaction and the most sustainable development model. This chapter repeats the adage that the destruction of archaeological sites has deterrent effects on the tourism industry. As this essay has demonstrated in its application of these concepts, there are a number of ways to avoid destruction through education and raising public awareness. In conclusion, I wish to emphasise that deliberate destruction should be punished. It is my hope that states and other stakeholders consider archaeological heritage a valuable and non-renewable asset worthy of protection and use, rather than a target of criminal behaviour (Hutchings & La Salle, 2017).

References

Al-Kanany, M.M.R. (in press) Extremist iconoclasm versus real Islamic values: Implications for heritage-based tourism development in Iraq. *Journal of Heritage Tourism*.

Ashworth, G.J. and Tunbridge, J.E. (2003) *Malta Makeover: Prospects for the Realignment of Heritage, Tourism and Development*. Groningen, Netherlands: University of Groningen, Urban and Regional Studies Institute.

Ashworth, G.J. and van der Aa, B.J.M. (2002) Bamyan: Whose heritage was it and what should we do about it? *Current Issues in Tourism* 5 (5), 447–457.

Bastian, F. and Alabouvette, C. (2009) Lights and shadows on the conservation of a rock art cave: The case of Lascaux Cave. *International Journal of Speleology* 38 (1), 55–60.

Bond, G.C. and Gilliam, A. (1994) *Social Construction of the Past: Representations as Power*. London: Routledge.
Butler, R.W. (2019) Tourism and conflict in the Middle East. In D.J. Timothy (ed.) *Routledge Handbook on Tourism in the Middle East and North Africa* (pp. 231–240). London: Routledge.
Bryant, J. (2016) Prehistoric cave art celebrated at new Lascaux centre in Dordogne. *The Guardian*. See https://www.theguardian.com/travel/2016/dec/15/prehistoric-cave-art-lascaux-dordogne-france-grotto-replica (accessed 15 August 2018).
Comer, D.C. (2012) *Tourism and Archaeological Heritage Management at Petra: Driver to Development or Destruction*. Dordrecht: Springer.
Crane, S.A. (ed.) (2000) *Museums and Memory*. Stanford: Stanford University Press.
De Cesari, C. (2015) Post-colonial ruins: Archaeologies of political violence and IS. *Anthropology Today* 31 (6), 22–26.
De Micheli, F. (2019) La Destruction Du Patrimoine: L'Eternelle Indignation. In A. Mouchtouris and F. De Micheli *Le Patrimoine De l'Autre: La Temporalité D'Une Injonction*. Paris: Le Manuscrit.
Gathercole, P. and Lowenthal, D. (eds) (1990) *The Politics of the Past*. London: Unwin Hyman.
Grégoire, A.H. (1793) *Rapport Sur les Destructions Opérées Par Le Vandalisme et Les Moyens De Le Réprimer*. séance du 14 fructidor, l'an second de la République une et indivisible, suivi du Décret de la Convention nationale. Paris: Imprimés et envoyés par ordre de la Convention nationale aux administrations et aux sociétés populaires. See https://donum.uliege.be/retrieve/19671 (accessed 21 July 2018).
Halbwachs, M. (1968) *La Mémoire Collective*. Paris: PUF.
Halbwachs, M. (1976) *Les Cadres Sociaux de la Mémoire*. Paris: Mouton.
Hole, B. (2013) A many-cornered thing: The role of heritage in Indian nation-building. *Journal of Intervention and Statebuilding* 7 (2), 196–222.
Hutchings, R.M. and La Salle, M. (2017) Archaeology as state heritage crime. *Archaeologies: Journal of the World Archaeological Congress* 13 (1), 66–87.
Jones, C.W. (2018) Understanding ISIS's destruction of antiquities as a rejection of nationalism. *Journal of Eastern Mediterranean Archaeology & Heritage Studies* 6 (1–2), 31–58.
Kirshenblatt-Gimblett, B. (1995) Theorizing heritage. *Ethnomusicology* 39 (3), 367–380.
Layton, R. and Thomas, J. (2001) Introduction: The destruction and conservation of cultural property. In R. Layton, P. Stone and J. Thomas (eds) *Destruction and Conservation of Cultural Property* (pp. 1–21). London: Routledge.
Lowenthal, D. (1985) *The Past is a Foreign Country*. Cambridge: Cambridge University Press.
Lowenthal, D. (1998) *The Heritage Crusade and the Spoils of History*. Cambridge: Cambridge University Press.
Lowenthal, D. (2015) *The Past is a Foreign Country, Revisited*. Cambridge: Cambridge University Press.
McGill, G. (2003) *Building on the Past: A Guide to the Archaeology and Development Process*. London: Taylor & Francis.
Meskell, L.M. (2005) Sites of violence: Terrorism, tourism, and heritage in the archaeological present. In L. Meskell and P. Pels (eds) *Embedding Ethics* (pp. 123–146). Oxford: Berg.
Meskell, L.M. (2015) Gridlock: UNESCO, global conflict and failed ambitions. *World Archaeology* 47 (2), 225–238.
Mitchell, T. (2002) *Rule of Experts*. Berkeley: University of California Press.
Nora, P. (ed.) (1984) *Les Lieux de Mémoire*. Paris: Gallimard.
Nora, P. (1989) Between memory and history: Les Lieux de Mémoire. *Representations* 26, 7–25.

Olsen, D.H. (2019) Religion, pilgrimage and tourism in the Middle East. In D.J. Timothy (ed.) *Routledge Handbook on Tourism in the Middle East and North Africa* (pp. 109–124). London: Routledge.

Pollini, J. (2013) The archaeology of destruction: Christians, images of antiquity, and some problems of interpretation. In S. Ralph (ed.) *The Archaeology of Violence: Interdisciplinary Approaches* (IEMA Proceedings 2) (pp. 241–267). Albany: State University Press of New York.

Sax, J.L. (1990) Heritage preservation as a public duty: The Abbé Grégoire and the origins of an idea. *Michigan Law Review* 88 (5), 1142–1169.

Shahab, S. and Isakhan, B. (2018) The ritualization of heritage destruction under the Islamic State. *Journal of Social Archaeology* 18 (2), 212–223.

Smith, C., Burke, H., de Leiuen, C. and Jackson, G. (2016) The Islamic State's symbolic war: Da'esh's socially mediated terrorism as a threat to cultural heritage. *Journal of Social Archaeology* 16 (2), 164–188.

Sørensen, M.L.S. and Carman, J. (eds) (2009) *Heritage Studies: Methods and Approaches* London: Routledge.

Srivastava, S. (1994) The abuse of history: A study of the white papers on Ayodhya. *Social Scientist* 22 (5/6), pp. 39–51.

Timothy, D.J. (2011) *Cultural Heritage and Tourism: An Introduction*. Bristol: Channel View Publications.

UNESCO (2015) #Unite4Heritage at the 38th UNESCO General Conference. See http://www.unesco.org/new/en/general-conference-38th/unite4heritage/ (accessed on 10 October 2018).

van Pinxteren, E. (2018) Beirut's make over – The meaning of 'Solidere' 15 tears later. See https://www.asfar.org.uk/beiruts-make-over-the-meaning-of-solidere-15-years-later/ (accessed 21 October 2018).

9 Plundering the Past: Tourism and the Illicit Trade in Archaeological Remains

Dallen J. Timothy

Introduction

Every year hundreds of millions of trips are taken across international borders for pleasure and relaxation, business, education and a wide variety of other reasons. One of the most common activities undertaken by travellers is shopping, with some evidence suggesting that it is in fact tourists' most common leisure pursuit in many destinations (Jansen-Verbeke, 1991, 1998; Timothy, 2005). Shopping can be a secondary attraction or activity in the destination, or it may be a primary motive for travel. While tourists shop for a wide array of merchandise, the most pervasive tourist retail items are souvenirs in many forms, including clothing, jewellery, electronics, alcohol, sweets, foodstuffs and handicrafts. Some tourists seek out 'place-based souvenirs' that are considered 'authentic' and illustrative of the destination, while others purchase whatever mementos are available (Swanson & Timothy, 2012).

In addition to traditional souvenirs such as wood carvings, baskets and pottery, coffee mugs, hats, T-shirts and textiles, there is a growing trend among some tourists, who may have a particular interest in archaeology or antiquities, to purchase ancient relics, whether authentic or not, during their journeys. There is also a growing trend wherein more and more collectors are travelling for the sole purpose of expanding their collections, which demonstrably contributes to the expanding illegal trade in antiquities throughout the world, particularly the developing world (Di Lernia, 2005; Timothy & Nyaupane, 2009). This chapter examines the global problem of illicit trade in archaeological relics (the term 'antiquities' will also be used interchangeably in this chapter) from a broad perspective, what fuels it, and the challenges associated with preventing it.

Trends in the illicit antiquities market are highlighted with these general patterns and problems brought into the realm of tourism to illustrate the role of tourism and how the global flow of travellers creates a unique demand for the trade in ancient relics, both legal and illegal. The chapter then focuses on the supply side of the antiquities trade in tourism and suggests seven primary relationships.

Tourism and the Antiquities Trade

Some observers suggest that people collect antiques and other memorabilia as a way of bolstering themselves by achieving realistic goals and concrete validation (McIntosh & Schmeichel, 2004) or because antiquities embody the experience of travel and, because people travel to experience the 'other', antiquities 'allude to the "otherness" of the distant past' (Evans-Pritchard, 1993: 11). Ancient artefacts might also be markers of cultural continuity, or a lost past, that have value as historical documents for collectors, museums and researchers, and are therefore worthy of being collected (Evans-Pritchard, 1993; Lobay, 2016). People have also long taken pieces of their destination as a way of 'proving' or documenting their visit. This was particularly important during medieval pilgrimages, where pilgrims often despoiled sacred sites by pocketing stones, leaves or cultural relics to display at home (Ron & Timothy, 2019).

Regardless of the underlying motives, which are undoubtedly many, there is a growing trend in the developed portions of the world of people collecting archaeological relics and other antiquities. This collecting trend manifests in increasing numbers of collectors and traders travelling for the purpose of buying and selling ancient artefacts – in some cases a form of business travel for dealers (discussed below) or leisure travel (for collectors) that has many links to growing trends in trade fairs and exhibitions.

While a few authors have examined the connections between tourism and modern antique collecting (Dutton & Busby, 2002; Grado *et al.*, 1997; Jones & Alderman, 2003; Loeb, 1989; Michael, 2002; Timothy, 2005), tourism-related discussions about the trade in ancient antiquities are still very scarce. Antiquities shopping tends to be concentrated in areas that have a distinctive and well-acknowledged heritage identity, and collectors are not a homogenous group. There may be significant differences in interests, habits and travel desires between individuals who prefer 'antiques' and those who collect 'antiquities'. In addition to the antiques that mark the recent past of a few centuries or less and as an extension of antiques and tourism, there is a booming trade in ancient artefacts, or antiquities, that typically hold a higher monetary value, are much older, are more challenging to acquire, and symbolise a much more antiquated heritage. These primordial remnants of human civilization, sometimes with an emphasis on a particular period or region, are growing in popularity among collectors and are the focus of this chapter.

As already noted, demand for archaeological relics among collectors worldwide is substantial. This includes museums. Estimates suggest that collectors number in the hundreds of millions, fuelling a thriving multi-billion dollar black market trade in ancient antiquities that continues to increase each year (Atwood, 2004; Bowman, 2008; Brodie, 2005; Campbell, 2013; Clarke & Szydlo, 2017; Dempsey, 1994; Mackenzie, 2002), lagging only behind drugs and armaments in the black market. Not all antiquities dealings are illegal, however, as there is a substantial legal antiquities trade that also accounts for billions of dollars each year.

The global antiquities trade is comprised of three components: the supply from source nations, the demand from the global market and the chain of supply and transportation in the middle (Mackenzie, 2002). The growing worldwide demand for relics has spurred an increase in grave robbing and lootings of archaeological sites. Demand for artefacts began in ancient times with travel for trade and global exploration. Important gold and silver figurines from faraway lands were traded for food, cookware, guns, paper and other novelty items from Europe and the Levant. During the colonial era, from the 15th to the 20th centuries, demand grew even more as the treasures of ancient civilizations were confiscated, plundered, or in many cases, purchased outright by colonial elites for transport to the homeland for display in personal collections and museums (Mueller, 2016; Ron & Timothy, 2019; Timothy & Boyd, 2003). Timothy (2005: 110) underscores that:

> Through time, as indigenous and outsider contact grew, demand for material icons of faraway and exotic places also grew, resulting in a thriving, albeit damaging, trade in antiquities and items of material culture. The illicit trade in valuable heritage heirlooms grew in many parts of the world, where graves and sacred sites were robbed and destroyed for their ancient riches, sunken ships were broken up and raided for their treasures, and statues, frescoes, and mosaics were pillaged from ancient temples and monuments, so that unseen collectors could augment their collections.

Sometimes, colonial administrators even claimed to be taking cultural artefacts away to 'protect and preserve' them for the good of the colony (Lowenthal, 2008; Mueller, 2016).

During the 19th and 20th centuries, more widespread travel from northern Europe to the Mediterranean and Asia saw many relics return to Europe from the 'other' world to enhance private collections and public displays (Tahan, 2017). The fabled history of places such as ancient Greece, the Roman Empire, the Holy Land, Turkey and Egypt, has put Mediterranean antiquities in high demand. The archaeological richness of, and inability to enforce antiquities laws in, countries such as Turkey, Egypt and several in Eastern Europe make their treasures a hot commodity in the European and North American antiquities marketplace (Ghanem & Saad, 2015; Özdoğan, 1998). In the words of Lowenthal (2008: 381), 'the global admiration that has cost the Mediterranean so many sites and stones, statues and paintings, leaves a storehouse still so overflowing that only a fraction of it is yet unearthed, and of that, a smaller fraction properly assessed, curated or displayed'. Similar trends are found throughout Asia and Latin America, where according to Mueller (2016: 64),

> From murderous temple thieves in India to church pillagers in Bolivia to hundred-man bands of tomb raiders in China's Lioaning Province, looters are strip-mining our past. Like most illegal activities, looting is hard to quantify. But satellite imagery, police seizures, and witness reports from the field all indicate that the trade in stolen treasures is booming around the world.

Another aspect of demand that fuels illegal trade is museum collections. For many years, museums have purchased antiquities from diggers in an attempt to curtail the diggers' trade abroad. This, however, has backfired in several ways, one of which is contributing to additional illegal excavations as grave robbers see the value that some museums place on artefacts and their willingness to pay for them (Clarfield, 2008). Mueller (2016: 65) notes that many museum curators, in their efforts to safeguard humankind's tangible past, 'rescue' antiquities from unstable countries 'even if it means buying from looters'. As Özdoğan (1998: 121) notes in the context of Turkey, 'to stop the illicit export of antiquities, buying them in Turkey (for Turkey) by paying sums comparable to the western collectors has been suggested as a solution. For some years, Turkish Museums bought from illicit diggers and, of course, only encouraged further destruction of the sites. Museums attained important objects at the expense of losing scientific knowledge of their contexts'. Clearly, not all museums purchase illegal artefacts. In fact, in recent years this has become a rarity, especially since the International Council of Museums has adopted a strict code of ethics that prohibits the importation, exportation and purchase of illegal antiquities.

These historical and contemporary trends in acquiring items from abroad have resulted in a current debate about the repatriation of relics from the countries that currently host them to the countries where the artefacts originated (Chiwara, 2019; Tahan, 2017). Origin countries argue that artefacts must be returned, while possessing states maintain that they are more qualified to protect the world's heritage and that these items have, through time, become important markers of their own national identities.

From a supply perspective, this problem is especially acute in the less-developed world where poor farmers and fishermen, in an effort to boost their meagre incomes, undertake clandestine excavations and plunder ancient tombs to scavenge for antiquities that can be sold to dealers and collectors, including tourists. In only a few days of unearthing, these 'subsistence diggers' can earn more money than working for months at a time in the fields. Two decades ago, Matsuda (1998: 91) estimated that nearly a million people in the Mexican states of Guerrero, Vera Cruz, Jalisco, Chiapas and the Yucatan Peninsula were engaged in supplemental digging. Some 500,000 people were estimated to dig part time in Guatemala and El Salvador, and upwards of 50,000 in Belize. Similar trends have been observed in China, Cambodia and many countries of Africa and Latin America (Ciochon & James, 1989; Kankpeyeng & DeCorse, 2004; Labi & Robinson, 2001; Meo, 2007; Mortensen, 2006; Zhang, 1992).

Unfortunately, in many regions of the world, the western notion of conservation for its scientific, educational and cultural value has not caught on, for in developing regions it is difficult to garner support for protecting archaeological artefacts when people's lives are at stake and when parents struggle to feed their families (Timothy, 1999; Timothy & Nyaupane, 2009). In fact, many indigenous societies see ancient burial places and their entombed treasures as gifts intentionally left to them by their ancestors for their modern-day good fortune and economic gain. As such, many aboriginal peoples feel justified in harvesting ancestral riches (Matsuda, 1998; Timothy & Boyd, 2006). While it is a small cohort, proponents of free trade in antiquities argue that it is beneficial for the local poor because it employs them, it helps preserve material culture in private and museum collections, and promotes appreciation for a wide range of art forms (Brodie, 2003).

Challenges to antiquities protection

Unfortunately, in many places, heritage protection legislation is too relaxed and was enacted too late to protect much of the tangible past. While the heritage of many countries, such as some under previous Ottoman control, which established its first antiquities laws in 1869, was protected relatively early, in numerous other colonial realms, national heritage protection laws were not passed until far into the 20th century, often only after independence was gained from European metropoles. Such was the case in several French and British territories of West Africa, for example. This has had far-reaching implications for antiquities protection throughout Africa, Asia, Latin America and the Middle East.

For this and other reasons, Egypt is one of the most plundered places in the world where, according to Mueller (2016: 64), approximately one quarter of all known archaeological areas have been damaged by looters and pillagers. Current legislation in that country, however, exacts strict punishments for people involved in illicit antiquities trading. Smugglers, for instance, face life imprisonment and a fine of up to 500,000 Egyptian pounds (USD $91,300) (Egyptian Cultural Heritage Organisation, 2008). There has been a surge of national laws in many countries that result in heavy penalties as regards illicit antiquities. In addition, several international agreements have also been ratified to curtail looting and illegal trade in archaeological heritage, including UNESCO's 1970 Convention on the Means of Prohibiting and Preventing the Illicit Import, Export, and Transfer of Ownership of Cultural Property, as well as the 1995 UNIDROIT Convention on Stolen or Illegally Exported Cultural Objects. According to Ghanem and Saad (2015), however, in Egypt there are many loopholes in Egyptian law that allow residents in archaeology-rich areas of the country to possess antiquities and carry out building projects, and thereby excavations, in archaeological areas.

Today, despite the existence of laws and international accords banning unlawful antiquities trafficking, growing worldwide affluence, increased global mobility, extended leisure time, improved technology, as warring factions selling antiquities on the black market have led to increased numbers of collectors and dealers, more willingness to spend and riskier attempts to obtain forbidden artefacts. National and multinational protective measures are difficult to enforce, as their implementation is at the mercy of each signatory state, and most states in the global community lack the means or strength to be able to enforce these policies (Henderson, 2009; Mueller, 2016). Thus, illegal buying and selling of national treasures and remnants of 'world heritage' continues in full force with relatively few infractions that result in fines or jail sentences (Brodie, 2003; Brodie *et al.*, 2001; Timothy *et al.*, 2009).

Most developing countries and states in political transition (e.g. Eastern Europe, Vietnam, Cambodia, China) lack human resources in sufficient measure to enforce national and international laws. Ribeiro (1990) recognized this problem in the context of India, which suffers from untrained staff, a dearth of security personnel at important archaeological sites, and lack of skilled experts who can enforce preservation legislation. In Eastern Europe, particularly Russia, Serbia, Ukraine, Moldova and Romania, legal controls are much weaker than in Western Europe regarding the digging and export of Viking and Roman artefacts. For instance, it is extremely difficult, nearly impossible, to acquire Viking artefacts from Denmark, Norway, Sweden or Iceland nowadays, but a search of online sellers reveals that Viking relics from Russia and Estonia are ubiquitous on the international market.

In many developing countries, archaeological sites and historic buildings are often occupied and utilised by families or entire villages (Chakravarti, 2008; Shoup, 1985; Timothy, 1999; Timothy & Boyd, 2003). People establish their homes and communities in or around these historic sites, which creates several problems. First, the habitation of archaeological sites results in considerable wear and tear. Second, inhabitants make daily use of archaeological resources taken from the site, including dismantling for building materials, or utilizing ancient artefacts as tools or household items. Third, once they realize the bounty associated with antiquities, they begin to collect artefacts for sale to collectors and dealers. Finally, animals are usually allowed to graze and roam freely in archaeological areas and historical landscapes, damaging artefacts and polluting ancient structures (Timothy & Nyaupane, 2009). All four of these problems have important bearings on the trade in artefacts and tourism.

Political instability is also a salient menace, because it feeds illegal trade in several ways. Ghanem and Saad (2015) contend that the deterioration of Egypt's security environment following the 2011 revolution has made it easier for thieves to loot archaeological sites and sell antiquities. In war-torn regions, such as Somalia and the Democratic Republic of the Congo, unpaid and hungry soldiers find whatever means they can to purchase food, including looting museums, plundering graves and stealing (Labi & Robinson, 2001). Instability may also result in warring factions and terrorist organizations selling archaeology to international middlemen in order to fund their warfare (Brodie & Sabrine, 2018; Losson, 2017; Mustafa, 2019; Pollock, 2016). Mohamed Atta, the ringleader behind the 11 September 2001, terrorist attacks in the United States, was said to have financed the assault in part by selling stolen antiquities (Clarfield, 2008). There is also evidence that the Islamic State (ISIS) funded part of its terrorist activities by looting ancient Mesopotamian artefacts from museums and archaeological sites (Brodie & Sabrine, 2018; Mueller, 2016). Likewise, the US invasion of Iraq in 2003 led to widespread plundering of Iraqi museums by local gangs, warring factions, and other profiteers as museums were abandoned or used as shelters for insurgent fighters (Bogdanos, 2005; Brodie, 2003, 2005). Fortunately, many of the stolen artefacts have been recovered, although many will likely never be returned. Also, the Khmer Rouge paramilitary of Cambodia notoriously occupied the temples of Angkor Wat during the 1970s–1990s, destroying many artefacts and historic structures.

Tourism perspectives

Against this background it is easy to grasp how tourism is involved in the illicit trade in antiquities as destination residents realize the potential income to be made from selling artefacts to tourists (Brodie *et al.*, 2001; Evans-Pritchard, 1993; Kankpeyeng & DeCorse, 2004; Pollock, 2016; Winter, 2006). It should be noted again, however, that not all antiquities sales are illegal. Many countries certify certain items for sale through authorized dealers. Nonetheless, there is an active movement around the world to stop allowing any degree of trade in antiquities. Holy Land antiquities are in particular demand throughout the Christian world (Kersel, 2014). These can be purchased on the internet, but visiting the Holy Land and purchasing them on site adds to the experience and makes the pieces more intrinsically valuable and meaningful.

As noted in the beginning of the chapter, tourist shopping is either a primary or motive for travel or an ancillary action that occurs while visiting a place for other reasons. The same can be seen in the context of antiquities trading. Many thousands of people travel each year in search of antiquities to augment their collections. They intentionally visit dealers in countries that have fewer regulations than others have or lack the ability to enforce extant laws. However, as in the context of shopping in general, impulse buying is salient in the context of antiquities, as some tourists are convinced to purchase antiquities once in the destination, many without ever having thought of doing this beforehand (Timothy, 2005). Many countries have strict controls over what sorts of artefacts can be exported, usually requiring certificates and permits, but even from these countries, ancient artefacts regularly disappear in people's luggage (Atwood, 2004). Some countries have few controls in place to monitor such activities, which encourages antiquities plundering even further.

From a supply perspective, several types of relationships between tourism and antiquities consumption can be identified (Table 9.1). All of these relationships between buyers and sellers can involve serious collectors who travel for the purpose of acquiring early pieces or casual consumers who are enticed into buying small objects in the destination.

First and perhaps most common in some locations, is residents selling artefacts to tourists as part of the informal economy. This happens frequently in the developing world and typically involves small-scale transactions on streets, in markets, in hotels or at tourist attractions. Regardless

Table 9.1 Supply-side relationships between tourism and antiquities consumption

	Relationship	Characteristics
1	Casual, small-scale sales to tourists in public places	Tourists are convinced to buy items from informal traders who approach them on site. These sellers are typically small-scale and are not supported by any individual shop.
2	Legally licensed antiquities shops selling to tourists	Shopkeepers are licensed to sell antiquities to tourists. They are approved and certified agents by government offices in charge of cultural protection.
3	Licensed dealers selling illegal artefacts	Licensed shops sometimes sell illegal pieces to supplement their income at a significant risk.
4	Dealers and brokers travel to purchase or sell their relics	Many people travel to source countries to buy from excavators or middle people and return home to sell their goods on the antiquities market.
5	Tourists themselves find ancient objects or conduct their own digs	Many tourists pocket artefacts they find at archaeological sites or they undertaken illegal digs themselves. This is very risky, and many tourists have been arrested and fined.
6	Tourist demand causes the development of fake antiquities (copies of ancient artefacts)	As supplies of authentic relics are depleted or as it makes good business sense, fake artworks and relics are made to satisfy tourist demand.
7	Tourists are ripped off by dodgy antiquities dealers	Unsuspecting tourists are commonly scammed into believing their purchases are real and authentic, but they have little recourse in this case.

of the strict antiquities laws in Jordan, Egypt and Turkey, it is common for tourists to be approached in those countries by destination residents with petitions to buy small trinkets or artefacts that the seller might have found on his or her farm, or acquired in some other way (Atwood, 2004; Shoup, 1985; Timothy & Boyd, 2006). The seller sees this as an opportunity to generate a few extra dollars; the tourist sees it as an authentic and unique souvenir that represents the historicity of the destination. Some tourists, even those who had not previously thought of buying an historic artefact, often cave to the sales pressure and purchase small and uncertified antiquities. According to one study, some 20% of visitors to Angkor Wat purchased antiquities, and most do not realize the gravity of what they are doing (Meo, 2007; O'Reilly, 2014). This points to the important role of tour guides, travel agencies and archaeologists in raising awareness about the criminal nature of dealing in antiquities, although this is not a common practice yet in most parts of the world. In Cambodia's capital, Phnom Penh, ancient souvenirs can be bought for only a few US dollars (Meo, 2007). Similarly, in Egypt many tourists come to archaeological sites near Luxor with the notion in mind to buy some genuine Egyptian artefacts. 'Peddlers hanging around the tombs may approach them or local guides may offer to show them items for sale in the privacy of their homes. The interaction between local dealer and prospective buyers has its own dynamics. Items offered for sale generally include both genuine artefacts and "modern antiques", the latter suitably altered with a range of treatments such that only an expert eye can recognize they are fake' (van der Spek, 2008: 167).

Second, in legal shops, ancient relics can be bought and sold lawfully. Regular antique shops are commonplace in destinations, but relatively few countries will allow ancient antiquities to be sold on the open market. Israel and Palestine are exceptions, and dozens of sellers in Jerusalem and Bethlehem have stocked their shelves with oil lamps, pots, coins, Roman or Hellenistic metal tokens, Roman glass, small statues, bronze arrowheads and many other artefacts (Figure 9.1). Some of these, such as Roman-era oil lamps or coins, have important Biblical connotations and are especially salient souvenirs for religious tourists. In common with other spiritual seekers in other destinations (Goss, 2004; Shackley, 2001, 2006; Timothy, 2008), Christian pilgrims often feel more closely connected to the Holy Land by taking home a souvenir from the approximate time of Christ and which was mentioned specifically in the Bible.

The third type is a crossover between the first two – legal and licensed dealers selling illicit artefacts. While in Israel it is illegal for licensed antiquities traders to sell inauthentic merchandise and prohibited items (which apparently are checked regularly by the Antiquities Authority), some dealers will risk losing their licenses, heavy fines and even jail time by offering clandestine relics in the shadows of their shops. This is not uncommon in many parts of the world, including Latin America, Asia, the Middle East

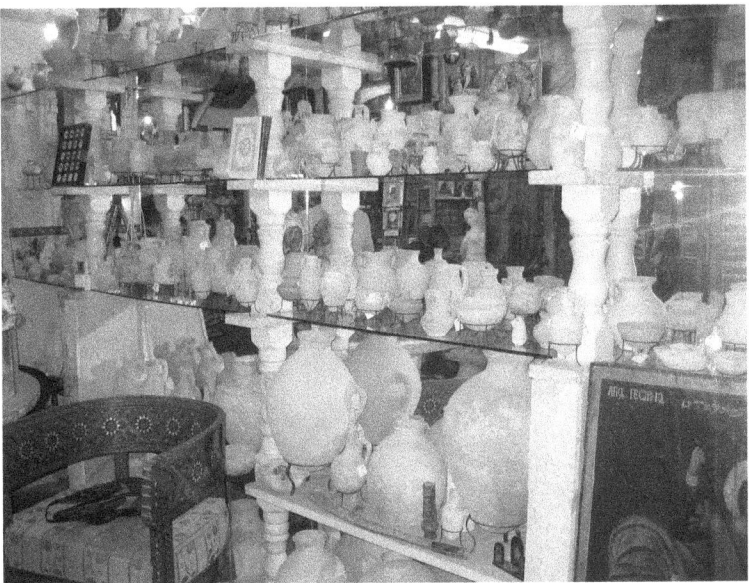

Figure 9.1 This antiquities shop in Jerusalem provides certified and legal sales of archaeological artefacts to tourists (Photo: Dallen J. Timothy)

and Africa (He, 2001; Mueller, 2016; Mugnai *et al.*, 2017). In some cases, a legitimate legal showroom functions as a front to a more surreptitious black market operation. Progler (1999) provides an example from Turkey, where sometimes tourism is used as a façade for the illegal trade in Ottoman-period antiquities. Progler (1999: 1) notes that

> although it is technically illegal to sell Ottoman antiquities in Turkey, the high prices they bring in the west make this a tempting and lucrative trade for unscrupulous buyers and sellers, who do not think twice from banking a fortune on the shards and scraps of the great Ottoman Islamic civilization. [One dealer] reflects on his trade. 'Japanese tourists are among our best customers … and they especially value the old calligraphic pieces' … Today [the dealer] is negotiating with a 'collector' from Britain, who is interested in some 'authentic' Islamic art, something a cut above the tourist fare … Not to disappoint, [the seller] produces three Anatolian Qur'ans, intact and whole, and an array of individual Quranic pages that have been removed from their original bindings and framed to be sold piecemeal … the collector then leans in close. 'Can you get something rare, something older, more ornate, more collectible?' [The dealer] then replies, 'yes, but for something like that I will need funds in advance, for up to US $10,000 …' When [the dealer] gets his money, he will likely do one of two things…He will either pay someone to steal a Qur'an from the waqf of an unsuspecting Anatolian village mosque, or he will pay a poor family to part with an heirloom. In either case, the customer gets his relic, [the dealer] gets his money, and the Islamic heritage of Turkey gets ripped off.

In many cases, it is difficult for authorities to ascertain whether or not the source of relics was legal or illegal, and few countries have sufficient monitoring networks in place to verify the origins of the products on offer. Some of the licensed sellers in the Holy Land, for example, are known to acquire artefacts through various channels that are excavated undercover by Palestinians in the West Bank, who in some areas face unemployment rates of nearly 40% and who are no longer allowed to work in Israel (Schulman, 2002). In the face of these political and economic crises, to survive, individuals, organized groups and even entire villages operate unauthorized excavations with shovels and backhoes looking for pottery, glass, oil lamps, coins, jewellery and statuettes. These items are excavated in areas outside the control of the Israel Antiquities Authority, making monitoring and regulating extremely difficult, even though many items end up on store shelves in Jerusalem. In the process of excavating, tombs are destroyed, bones scattered, and the names of the deceased, which are sometimes inscribed on the burial boxes, are wiped out (Israel Antiquities Authority, 2009).

Another tourism link is antiquities dealers, brokers and traders travelling in search of products to buy and sell. Profit making is the most crucial variable in the antiquities trade, which is supported by clandestine networks, privacy codes, and dealer reputations. With a multi-billion dollar per year 'industry' and hundreds of thousands of dealers involved in the antiquities trade, the level of international and domestic travel must be immense, although as noted earlier, numbers cannot be verified, as there is no single international body that monitors these activities systematically. Local and regional agents purchase artefacts from looters, who then sell to international traders, who eventually sell to the brokers, who then sell directly to collectors and museums. This supply chain, which varies from place to place and reflects regional differences, involves a great deal of international and regional travel, stimulating a form of business tourism based on furtive illicit activities (Gallagher, 2017).

The fifth issue is tourists collecting artefacts themselves. While it is prohibited in nearly every part of the world to pick up pottery shards or other remains of material culture at archaeological sites, this is a universal problem among tourists (Di Lernia, 2005; McGinn, 2008). This action is partly a result of a 'collectors' syndrome' that drives people to want to possess a piece of a significant place they have visited (Swanson & Timothy, 2012; Timothy, 2011). Some tourists might even go to extremes of digging at archaeological sites themselves after dark, which was also evident in the 19th century when travellers and diplomats would dig late at night and send their finds to curators of international museums. In fact, the flourishing 'treasure hunting tourism' sector encourages tourists to scavenge, with or without metal detectors (Rasmussen, 2014; Thomas, 2016; Thomas *et al.*, 2016). Some of the promoted activities are of dubious legality.

Tourists' collecting behaviour can lead to arrests, prison sentences, and hefty fines. Many examples of tourists being imprisoned for stealing or trying to export illegal antiquities abound in the media. Two examples include a Canadian high school student visiting Greece in 2005. On a path near the Parthenon, she picked up a rock and was immediately arrested by Greek police. Her case was dismissed a few days later when a judge believed her story that she was planning to use the rock only as a prop in a photograph. In April 2008, a Finnish tourist was arrested for trying to chip off a portion of an earlobe of an ancient moai statue on Easter Island. He was required to pay a $17,000 fine and write a public apology, and he was prohibited from returning to the island for three years (McGinn, 2008).

The sixth link (see Table 9.1) is the development of fake antiquities that become mass produced as tourist art. This becomes a negative issue when these touristic pieces are sold deceptively as authentic art works, and when they are mass produced without any kind of cultural context or meaning. In this process, the value and spiritual or utilitarian connection with the artefact is lost (Cohen, 1993; Graburn, 1984). Even when the artefacts' fakeness is a known fact, it is still a highly contentious issue among many cultural studies specialists, because it masquerades inauthentic relics as authentic and often results in the perpetuation of cheap 'tourist trash', 'tourist kitsch' or 'airport art' that is far separated from the cultural roots it tries to portray and the deep cultural meanings associated with the originals (Cohen, 1988; O'Connor, 2006; Rowlands, 2002).

Van der Spek (2008) provides an interesting case of an Egyptian village (al-Qurna), which is set inside ancient burial grounds at the Theban Necropolis. Residents of the village have long engaged in tomb robbing and selling their finds to tourists and international dealers until the resources began to diminish and authorities began to crack down. As a result, the villagers (Qurnawi) began to concoct counterfeit antiquities for sale to tourists and collectors. Even before the arrival of tourists, the villagers settled in that location intentionally to exploit the area's rich archaeology, and they have been involved in digging, making and selling antiquities ever since. They have become quite skilled at creating phoney artefacts, and many people in the village are employed in this dubious trade. In the process of creating fake relics, however, they destroy many of the old ones. For instance, Van der Spek (2008: 166) cites the practice of smashing original, whole papyri and using the fragments to coat the outside of forged ones to make them appear authentic. Similar occurrences have been well documented in Belize, Peru, Nepal, Nigeria, Jordan and other countries where the supply of desirable relics has dried up and villagers, desperate to continue their livelihoods, turn to manufacturing and selling replicas (Evans-Pritchard, 1993; Tarawneh & Wray, 2017).

Similar to the last issue is that countless tourists are ripped off each year, paying premium prices for what they believe to be 'authentic', or original, ancient works of art. One estimate suggests that approximately 90% of all African antiquities entering the US for and with collectors are replicas made to look ancient (Labi & Robinson, 2001). Many unsuspecting tourists fall prey to unscrupulous vendors who claim to be selling authentic artefacts but which are in reality modern fakes. In the antiquities trade, there is little legal recourse for such an occurrence, given that the collection and purchase of the historic pieces were probably of dubious legality anyway. In addition, sellers are often of the mindset that tourists are leaving and will not be coming back, so that even if they find out a piece is fake, they will be unable to return it or complain about it in person.

Conclusion

The illegal trade in antiquities is widespread and growing at an enormous rate. In some tourist destinations, selling antiquities is a legal activity, but even legalized antiquities sales have dubious elements about them, and some activists seek to ban all forms of trade in archaeological relics. In 1999, UNESCO published a code of ethics for dealers of cultural property. For the most part, it emphasizes not importing, exporting or transferring ownership of antiquities whenever there is any question that they might have been excavated illegally. UNESCO (2018) has also produced videos and advertisements recently for distribution to major airports as a means of raising awareness about not buying cultural relics.

At the root of the antiquities trade are poverty and greed – poverty among those who dig and receive a small pittance for their efforts, which if discovered could land them years in jail and thousands of dollars in fines, and greed among dealers and collectors who take advantage of the destitute populations of the developing world. Regarding this problem, Progler (1999: 3) argues that,

> the illegal trade in Qur'ans and other antiquities in Turkey is symptomatic of broader political and economic weaknesses, and reveals yet another insult to Islamic civilization by the west, which thinks it can buy, lie, steal, and brag its way into even the most sacred corners of Muslim history and culture. It is dangerous to ignore these examples of western imperialism, as it is precisely such creeping and insidious aspects of neo-colonialism, and their concomitant industries like tourism, that often do the most irreparable and lasting damage.

Tourism is a significant culprit in the sustained growth of this illicit activity, and it seems that whenever tourism is linked to archaeology, it results in stealing and the reproduction of antiquities. Collectors throughout history have plundered ancient sites and museums, and the modern era is no exception. Collectors, dealers and brokers still travel the world searching for new finds to augment their collections or to sell on the black market. From the perspective of tourists, even though archaeological museums may be interesting and appealing, 'the antiquity that can be purchased is a far more influential symbol for the average person than a similar object in a museum case' (Evans-Pritchard, 1993: 17). Some tourists themselves even engage in stealing from, or plundering, archaeological sites, a common practice known to residents in the local communities. In response to this 'tourist' demand, farmers and other residents target tourists for their propensity to want a piece of the destination they visit. Likewise, even legally certified antiquities dealers will conduct clandestine business under the guise of their legal license, and growing demand leads to the production of fake antiquities that are sold to tourists as authentic pieces.

Unfortunately, the growing demand for international travel and the demonstrated constant supply of antiquities available to tourists has fuelled the looting and destruction of archaeological sites. While this is most evident in the developing world, it is still a salient problem in Western Europe and North America as well. While scholars have acknowledged many of the negative ecological and social impacts of tourism, this destructive aspect of the relationship between archaeology and tourism has been ignored almost entirely by scholars of archaeology, tourism and the antiquities trade. It is incumbent upon researchers to begin examining this phenomenon in more depth in more locations to understand the situation better from political, economic, geographic, social, conservation and legal perspectives, and discover solutions to the problem.

More efforts are needed to raise awareness among tourists that archaeological remains are to be 'consumed' only symbolically for knowledge and learning, not to be vandalized or acquired as souvenirs to put on display at home. Governments must increase their efforts to train civil servants and heritage stewards in relevant antiquities departments to enforce the laws better and to help ensure the protection of archaeological remains in other ways. UNESCO's (2018) recent publication, *Fighting the Illicit Trafficking of Cultural Property*, in collaboration with the European Union, is a step in the right direction and should become a valuable training tool as countries figure out the best way of fighting the illegal trade in antiquities.

Unlike many elements of the natural environment, ancient artefacts of material culture are a non-renewable resource. The past has gone, and once these relics are gone and their contexts destroyed by plunderers and tourists, the knowledge contained within them will be lost forever.

References

Atwood, R. (2004) *Stealing History: Tomb Raiders, Smugglers, and the Looting of the Ancient World*. New York: St. Martin's Press.

Bogdanos, M. (2005) The casualties of war: The truth about the Iraq Museum. *American Journal of Archaeology* 109 (3), 477–526.

Bowman, B.A. (2008) Transnational crimes against culture: Looting at archaeological sites and the 'grey' market in antiquities. *Journal of Contemporary Criminal Justice* 24 (3), 225–242.

Brodie, N. (2003) Stolen history: Looting and illicit trade. *Museum International* 55 (3/4), 10–22.

Brodie, N. (2005) Illicit antiquities: The theft of culture. In G. Corsane (ed.) *Heritage, Museums and Galleries: An Introductory Reader* (pp. 122–140). London: Routledge.

Brodie, N., Doole, J. and Renfrew, C. (eds) (2001) *Trade in Illicit Antiquities: The Destruction of the World's Archaeological Heritage*. Cambridge: McDonald Institute.

Brodie, N. and Sabrine, I. (2018) The illegal excavation and trade of Syrian cultural objects: A view from the ground. *Journal of Field Archaeology* 43 (1), 74–84.

Campbell, P.B. (2013) The illicit antiquities trade as a transnational criminal network: Characterizing and anticipating trafficking of cultural heritage. *International Journal of Cultural Property* 20 (2), 113–153.

Chakravarti, I. (2008) Heritage tourism and community participation: A case study of the Sindhudurg Fort, India. In B. Prideaux, D.J. Timothy and K.S. Chon (eds) *Cultural and Heritage Tourism in Asia and the Pacific* (pp. 189–202). London: Routledge.

Chiwara, D. (2019) Documentation: A security tool for the identification and repatriation of illicitly trafficked objects from museums with particular reference to the National Gallery of Zimbabwe. *Heritage* 2 (1), 390–399.

Clarfield, G. (2008) Stop the appeasement of art and antiquities thieves: Museums give criminals incentives to steal and steal again by paying ransoms for art works. *The Globe and Mail* (Toronto) 5 July, A19.

Clarke, C.M. and Szydlo, E.J. (2017) *Stealing History: Art Theft, Looting, and Other Crimes against Our Cultural Heritage*. Lanham, MD: Rowman and Littlefield.

Cohen, E. (1988) Authenticity and commoditization in tourism. *Annals of Tourism Research* 15, 371–386.

Cohen, E. (1993) The heterogeneization of a tourist art. *Annals of Tourism Research* 20, 138–163.

Ciochon, R. and James, J. (1989) The battle of Angkor Wat. *New Scientist* 124, 52–57.

Dempsey, M. (1994) Protectors of Peru's shining past. *New Scientist* 143, 23–25

Di Lernia, S. (2005) Incoming tourism, outgoing culture: Tourism, development and cultural heritage in the Libyan Sahara. *The Journal of North African Studies* 10 (3/4), 441–457.

Dutton, S. and Busby, G. (2002) Antiques-based tourism: Our common heritage? *Acta Turistica* 14 (2), 97–119.

Egyptian Cultural Heritage Organisation (2008) Illegal antiquities trade. *ECHO News and Articles*. See www.e-c-h-o.org/News/IllegalAntiquities/htm (accessed 10 September 2008).

Evans-Pritchard, D. (1993) Ancient art in modern context. *Annals of Tourism Research* 29 (1), 9–31.

Gallagher, S. (2017) 'Purchased in Hong Kong': Is Hong Kong the best place to buy stolen or looted antiquities? *International Journal of Cultural Property* 24 (4), 479–496.

Ghanem, M.M. and Saad, S.K. (2015) Enhancing sustainable heritage tourism in Egypt: Challenges and framework of action. *Journal of Heritage Tourism* 10 (4), 357–377.

Goss, J. (2004) The souvenir: Conceptualization of the object(s) of tourist consumption. In A.A. Lew, C.M. Hall and A.M. Williams (eds) *A Companion to Tourism* (pp. 327–336). Oxford: Blackwell.

Graburn, N. (1984) The evolution of tourist arts. *Annals of Tourism Research* 11 (3), 393–419.

Grado, S.C., Strauss, C.H. and Lord, B.E. (1997) Antiquing as a tourism recreational activity in southwestern Pennsylvania. *Journal of Travel Research* 35 (3), 52–56.

He, S. (2001) Illicit excavation in contemporary China. In N. Brodie, J. Doole and C. Renfrew (eds) *Trade in Illicit Antiquities: The Destruction of the World's Archaeological Heritage* (pp. 19–24). Cambridge: McDonald Institute.

Henderson, J.C. (2009) The meanings, marketing, and management of heritage tourism in Southeast Asia. In D.J. Timothy and G.P. Nyaupane (eds) *Cultural Heritage and Tourism in the Developing World: A Regional Perspective* (pp. 73–92). London: Routledge.

Israel Antiquities Authority (2009) Preventing Antiquities Robbery. See http://www.antiquities.org.il/shod_eng.asp (accessed 15 January 2010).

Jansen-Verbeke, M. (1991) Leisure shopping: A magic concept for the tourism industry? *Tourism Management* 12, 9–14.

Jansen-Verbeke, M. (1998) The synergism between shopping and tourism. In W.F. Theobold (ed.) *Global Tourism* (2nd edn) (pp. 428–446). Oxford: Butterworth-Heinemann.

Jones, K.L. and Alderman, D.H. (2003) Antiques tourism and the selling of heritage in eastern North Carolina. *North Carolina Geographer* 11, 74–87.

Kankpeyeng, B.W. and DeCorse, C.R. (2004) Ghana's vanishing past: Development, antiquities, and the destruction of the archaeological record. *African Archaeological Review* 21 (2), 89–128.

Kersel, M.M. (2014) The lure of the artefact? The effects of acquiring eastern Mediterranean material culture. In A.B. Knapp and P. van Dommelen (eds) *The Cambridge Prehistory of the Bronze and Iron Age Mediterranean* (pp. 367–378). Cambridge: Cambridge University Press.

Labi, A. and Robinson, S. (2001) Looting Africa. *Time* 158 (4), n.p.

Lobay, G. (2016) Looting and the antiquities trade. In S. Bell and A.A. Carpino (eds) *A Companion to the Etruscans* (pp. 458–473). Chichester: Wiley.

Loeb, L.D. (1989) Creating antiques for fun and profit: Encounters between Iranian Jewish merchants and touring coreligionists. In V.L. Smith (ed.) *Hosts and Guests: The Anthropology of Tourism* (pp. 237–245). Philadelphia: University of Pennsylvania Press.

Losson, P. (2017) Does the international trafficking of cultural heritage really fuel military conflicts? *Studies in Conflict & Terrorism* 40 (6), 484–495.

Lowenthal, D. (2008) Mediterranean heritage: Ancient marvel, modern millstone. *Nations and Nationalism* 14 (2), 369–392.

Mackenzie, S. (2002) *Regulating the Market in Illicit Antiquities*. Canberra: Australian Institute of Criminology.

Matsuda, D. (1998) The ethics of archaeology, subsistence digging, and artifact looting in Latin America: point muted counterpoint. *International Journal of Cultural Property* 7, 87–97.

McGinn, D. (2008) How to avoid a $17,000 souvenir. *The Globe and Mail* (Toronto) 04 June, R8.

McIntosh, W. and Schmeichel, B. (2004) Collectors and collecting: a social psychological perspective. *Leisure Sciences* 26 (1), 85–97.

Meo, N. (2007) Looting of relics is grave news for future generations. *South China Morning Post* 11 August, 14.

Michael, E. (2002) Antiques and tourism in Australia. *Tourism Management* 23 (2), 117–125.

Mortensen, L. (2006) Structural complexity and social conflict: Managing the past at Copan, Honduras. In D. Brodie (ed.) *Archaeology, Cultural Heritage and the Antiquities Trade* (pp. 258–269). Gainesville: University Press of Florida.

Mueller, T. (2016) Plundering the past: The illegal antiquities trade is booming, wreaking havoc on the world's archaeological heritage. *National Geographic* 229 (6), 58–81.

Mugnai, N., Nikolaus, J., Mattingly, D. and Walker, S. (2017) Libyan antiquities at risk: Protecting portable cultural heritage. *Libyan Studies* 48, 11–21.

Mustafa, M.H. (2019) Intangible heritage and cultural protection in the Middle East. In D.J. Timothy (ed.) *Routledge Handbook on Tourism in the Middle East and North Africa* (pp. 57–70). London: Routledge.

O'Connor, K. (2006) Kitsch, tourist art, and the little grass shack in Hawaii. *Home Cultures* 3 (3), 251–271.

O'Reilly, D.J. (2014) Heritage and development: Lessons from Cambodia. *Public Archaeology* 13 (1–3), 200–212.

Özdoğan, M. (1998) Ideology and archaeology in Turkey. In L. Meskell (ed.) *Archaeology Under Fire: Nationalism, Politics, and Heritage in the Eastern Mediterranean and Middle East* (pp. 111–123). London: Routledge.

Pollock, S. (2016) Archaeology and contemporary warfare. *Annual Review of Anthropology* 45, 215–231.

Progler, Y. (1999) Western 'collectors' fuel illegal antiquities trade in Turkey. Muslimedia, January 16. See www.muslimedia.com/archives/special99/tur-tourist3.htm (accessed 4 September 2008).

Rasmussen, J.M. (2014) Securing cultural heritage objects and fencing stolen goods? A case study on museums and metal detecting in Norway. *Norwegian Archaeological Review* 47 (1), 83–107.

Ribeiro, E.F.N. (1990) The existing and emerging framework for heritage conservation in India. *Third World Planning Review* 12 (4), 338–343.

Ron, A.S. and Timothy, D.J. (2019) *Contemporary Christian Travel: Pilgrimage, Practice and Place.* Bristol: Channel View Publications.

Rowlands, M. (2002) Heritage and cultural property. In V. Buchli (ed.) *The Material Culture Reader* (pp. 105–114). Oxford: Berg.

Schulman, M. (2002) Rise in antiquities theft vexes Israel's 'Indiana Joneses'. *Christian Science Monitor* 14 November, 13.

Shackley, M. (2001) *Managing Sacred Sites.* London: Continuum.

Shackley, M. (2006) Empty bottles at sacred sites: Religious retailing at Ireland's national shrine. In D.J. Timothy and D.H. Olsen (eds) *Tourism, Religion and Spiritual Journeys* (pp. 94–103). London: Routledge.

Shoup, J. (1985) The impact of tourism on the Bedouin of Petra. *Middle East Journal* 39 (2), 277–291.

Swanson, K.K. and Timothy, D.J. (2012) Souvenirs: Icons of meaning, commercialization, and commoditization. *Tourism Management* 33 (3), 489–499.

Tahan, L.G. (2017) Trafficked Lebanese antiquities: Can they be repatriated from European museums? *Journal of Eastern Mediterranean Archaeology and Heritage Studies* 5 (1), 27–35.

Tarawneh, M.B. and Wray, M. (2017) Incorporating Neolithic villages at Petra, Jordan: An integrated approach to sustainable tourism. *Journal of Heritage Tourism* 12 (2), 155–171.

Thomas, S. (2016) The future of studying hobbyist metal detecting in Europe: A call for a transnational approach. *Open Archaeology* 2 (1), 140–149.

Thomas, S., Seitsonen, O. and Herva, V.P. (2016) Nazi memorabilia, dark heritage and treasure hunting as 'alternative' tourism: Understanding the fascination with the material remains of World War II in northern Finland. *Journal of Field Archaeology* 41 (3), 331–343.

Timothy, D.J. (1999) Built heritage, tourism and conservation in developing countries: challenges and opportunities. *Journal of Tourism* 4, 5–17.

Timothy, D.J. (2005) *Shopping Tourism, Retailing and Leisure.* Clevedon: Channel View Publications.

Timothy, D.J. (2008) Genealogical mobility: Tourism and the search for a personal past. In D.J. Timothy and J.K. Guelke (eds) *Geography and Genealogy: Locating Personal Pasts* (pp. 115–135). Aldershot: Ashgate.

Timothy, D.J. (2011) *Cultural Heritage and Tourism: An Introduction*. Bristol: Channel View Publications.
Timothy, D.J. and Boyd, S.W. (2003) *Heritage Tourism*. Harlow: Prentice Hall.
Timothy, D.J. and Boyd, S.W. (2006) Heritage tourism in the 21st century: Valued traditions and new perspectives. *Journal of Heritage Tourism* 1 (1), 1–16.
Timothy, D.J. and Nyaupane, G.P. (2009) *Cultural Heritage and Tourism in the Developing World: A Regional Perspective*. London: Routledge.
Timothy, D.J., Wu, B. and Luvsandavaajav, O. (2009) Heritage and tourism in East Asia's developing nations: Communist-socialist legacies and diverse cultural landscapes. In D.J. Timothy and G.P. Nyaupane (eds) *Cultural Heritage and Tourism in the Developing World: A Regional Perspective* (pp. 93–108). London: Routledge.
UNESCO (1999) International Code of Ethics for Dealers in Cultural Property. See http://www.unesco.org/new/en/culture/themes/illicit-trafficking-of-cultural-property/legal-and-practical-instruments/unesco-international-code-of-ethics-for-dealers-in-cultural-property/ (accessed 3 March 2019).
UNESCO (2018) *Fighting the Illicit Trafficking of Cultural Property*. Paris: UNESCO.
van der Spek, K. (2008) Faked *antikas* and 'modern antiques'. *Journal of Social Archaeology* 8 (2), 163–189.
Winter, T. (2006) Ruining the dream? The challenge of tourism at Angkor, Cambodia. In K. Meethan, A. Anderson and S. Miles (eds) *Tourism Consumption and Representation: Narratives of Place and Self* (pp. 46–66). Wallingford: CABI.
Zhang, D. (1992) Protecting China's rich heritage of cultural relics. *China Today* 41 (6), 14–17.

10 Protecting the Archaeological Past in the Face of Tourism Demand

Jennifer P. Mathews

Introduction

Archaeologists have mixed feelings toward the development of archaeological heritage as a tourism resource. On the one hand, we recognize that local peoples have a need for, and a right to, economic development, and archaeological sites can provide a framework for that development. On the other hand, most of us are disheartened by the commodification of archaeological resources and the inevitable damage and destruction that occurs in the face of tourism (see for example, Díaz-Andreu, 2013; Gould, 2017). We also have to recognize that although as archaeologists we have had a privileged status in the protection of archaeological heritage, we can no longer deny that we are only one of the many stakeholders who have legitimate interests in this regard (Gould, 2017; Pacifico & Vogel, 2012). As Walker and Carr (2013: 14) state, 'Archaeologists do not work in isolation, and an archaeological site is not merely a pile of ruins or a collection of ancient things. Rather, there is a more complex setting that includes the present-day ecosystem, local communities, and the sociopolitical networks that must also be considered.' This chapter provides an overview of the recent research on the protection and conservation of touristic archaeological heritage through the lens of economic, cultural and ethical considerations.

The Evolution of Archaeology-based Heritage Tourism

Archaeological tourism began in earnest in the 15th century when early travelers visited ancient Egypt for religious pilgrimages and to see the pyramids (Pacifico & Vogel, 2012). In the 18th and 19th centuries, it was primarily the rich who traveled to visit spectacular cultural heritage, such as castles, forts, cathedrals and palaces, or who collected or donated master artworks as representatives of a particular period (Díaz-Andreu, 2013; Hassan, 2017). Archaeologists like Gertrude Bell began their careers as tourists following visits to Greece or the Middle East. After learning Arabic and spending extensive time in Iraq, Bell became Honorary Director of Antiquities, supervised excavations and founded the Baghdad Archaeological Museum (Howell, 2006). Tourists visited the excavations of Sir Williams Matthew Petrie in Egypt and Palestine, and he wrote books encouraging foreigners to visit countries such as Egypt (Mairs & Muratov, 2015). Archaeologist John Lloyd Stephens (1969) also inspired archaeological tourism through his 19th-century book series *Incidents of Travel*, which highlighted his adventures in Central America, Mexico and Egypt. Beautifully illustrated by artist Frederick Catherwood, these books brought Maya sites, such as Copán in Honduras and Chichén Itzá in the Yucatán, into the public imagination.

During the 20th century, with the introduction of wide-body jets and coach seating options, tourism incorporated the middle class with disposable income in search of a holiday, often with a focus on 'sun, sea and sand' (Díaz-Andreu, 2013; Gould, 2017; Pacifico & Vogel, 2012). However, with increased access to learning about archaeology and the cultural past through television and other popular media (Sabloff, 1998), visiting archaeological heritage sites has become one of the fastest growing sectors of the tourism industry (Díaz-Andreu, 2013; du Cros & McKercher, 2015). Since the 1970s, cultural tourism, in which visitors seek to gain a deeper understanding of the culture or heritage of a destination by visiting cultural attractions and events, has become a conventional form of travel (du Cros & McKercher, 2015; Timothy & Boyd, 2003). In fact, by 2012, the World Bank estimated that one out of every seven people on the planet had visited another country (Gould, 2017), and that between 35 and 80% of tourists are cultural tourists (du Cros & McKercher, 2015: 3).

And yet, as McKercher first noted in 2002, the cultural tourist market is not homogenous by any means, and cultural heritage serves different functions for visitors. A 'purposeful heritage tourist' has cultural heritage as a primary motivation for visiting a destination, and might be considered a more traditional visitor to an archaeological site. These visitors may come with background knowledge about a culture and are willing to explore areas with little tourism infrastructure (Alazaizeh *et al.*, 2016b; du Cros & McKercher, 2015; Wager, 1995). They might bring specialized books written about an archaeological site and be willing to rent a car and explore on their own. These individuals may also be aware of the significance of heritage protection, and are prepared to pay high entrance fees to subsidize conservation efforts (Wager, 1995). A 'sightseeing heritage tourist's' main goal is to learn and experience heritage but is more motivated by entertainment. These are the kinds of visitors who often have an expectation of tourist shops, restaurants and visitor centers as part of their experience. For the 'casual heritage tourist', the cultural heritage available in the destination has little impact on their travel choices, while 'incidental heritage tourists' are not enthused about heritage per se; they happen upon cultural sites during their travels. For example, incidental visitors might be motivated by photography and choose to visit an archaeological site because of its picturesque setting rather than its culture history. Finally, a 'serendipitous heritage tourist', although not driven by cultural tourism initially, happens upon heritage sites and ends up having a deep experience (Alazaizeh et al., 2016b; du Cros & McKercher, 2015; McKercher, 2002).

This means that most tourists see cultural heritage sites from a very different perspective than archaeologists do, and much to archaeologists' disappointment, they usually have a shallow sightseeing experience. If they have any interest in archaeology, they are often seeking a 'high novelty value' that allows them to feel a connection with the past that may be inauthentic from the perspective of the archaeologist (Walker & Carr, 2013). This also means that touristic archaeological sites now incorporate aspects that are far outside of most archaeologists' interests and expertise (Walker & Carr, 2013), including economic considerations, which will be discussed below.

Economics and Archaeological Heritage Tourism

Despite the general discomfort with the notion, it is important for archaeologists to recognize the economic potential that archaeological tourism can bring, which was the focus of Chapter 3. For many countries, it is one of the most profitable sectors of the economy and can generate billions of dollars annually (Cayron, 2017), and it has the potential to have a significant impact on local, national and international economies (Walker & Carr, 2013). Today, governmental and corporate policies have increasingly taken archaeological heritage into consideration in their economic impact statements (Gould, 2017). Because of the tremendous financial gains heritage tourism can have on local economies, government officials may view archaeological heritage sites as key revenue generators. However, this does not mean that these revenues will be devoted toward archaeological sites. In places like China, despite high ticket sales, little is reinvested into heritage conservation. For example, of all revenue generated at the Huangshan Scenic Area, only 1% was budgeted for conservation. Additionally, local governments may hope to increase the number of visitors and maximize revenues, with little consideration for their impact. From the archaeological perspective, this only increases the strain placed on the sites and escalates the need for heritage conservation (Díaz-Andreu, 2013; Shepherd & Yu, 2013).

When archaeological sites enjoy increased popularity, they experience dramatic wear and tear. For example, at the site of Petra, Jordan, soaring tourism has led to vehicular traffic breaking and damaging monuments and carvings, tourists wearing-away inscriptions and carving marks or toppling walls by sitting upon them, and donkeys' hooves damaging ancient steps while transporting tourists over them (Comer, 2012). Large numbers of visitors can also bump up against surfaces, increase humidity and introduce microorganisms and dust to internal sites, causing damage to murals, mosaics, wall paintings and artifacts. Tourists also cause an increase in vehicle traffic (in the form of tour buses, taxis and rental cars) and thus pollution, and may even engage in vandalizing sites (Orbaşli & Woodward, 2012). Unfortunately, the damage that occurs with high tourism traffic can be irreversible. Although the plan for managing heritage tourism should be put into place *before* bringing large numbers of visitors (Comer, 2012), too often this happens after the fact. For example, at Petra, as the result of site damage, the Jordanian Department of Antiquities and the Ministry of Tourism have increased the entrance fee for foreign tourists to 50 Jordanian Dinars (in comparison to 1 Jordanian Dinar for locals and Jordanian residents) to be invested in site conservation.

Proactive economic planning needs to assess heritage localities for their capacity for the number of visitors a site can hold at any one time (physical capacity), while also taking into consideration how much the natural and archaeological environment can withstand before damage occurs to the local ecology and archaeological features (environmental capacity). Tourism developers should also consider the quality of the visitor experience (perceptual capacity or social carrying capacity). Large crowds of visitors can dampen the ambiance, reducing a visitor's ability to enjoy a place or appreciate its beauty. Ultimately, this may discourage people from making return trips or recommending it to others (Alazaizeh *et al.*, 2016a; Orbaşli & Woodward, 2012; Wager, 1995). In particular, at locations like Angkor Wat in Cambodia, tourists may have an expectation of experiencing the spirituality and serenity of the World Heritage Site as part of their visit. While the total capacity of visitors can be increased with improved infrastructure and management, this needs to be a mindful process. Although predicting tourist demand can be tricky, it may be conducted by making comparisons with the number of visitors to other cultural sites within the region (Wager, 1995). Further, not all heritage tourists will have the same benchmarks for what is considered overcrowding, and thus it is important to obtain information from visitors about these expectations when designing management standards (Alazaizeh *et al.*, 2016a).

New development for tourist accommodations should be thoughtfully planned so that it is concentrated into zones that minimize damage to the local environment and archaeological resources, allows services like roads, electricity and water to be shared, and thus provides mutual benefits for local communities and tourists alike (Loulanski & Loulanski, 2011; Wager, 1995). Although limits on development have traditionally been seen as limits on economic benefits, there is now some awareness that protecting these resources is the only sustainable way to maintain economic development (Wager, 1995).

Nonetheless, some developers and government agencies have placed the economic development around heritage places ahead of protecting the sites (Gould, 2017). For example, the Four Seasons Hotel has been given permission to build over a medieval palace in Istanbul, despite the palace's World Heritage status (Orbaşli & Woodward, 2012). Similarly, Liverpool's Maritime Mercantile City World Heritage Site in England was created to protect the 18th-, 19th- and early 20th-century cultural heritage sites, while promoting job growth and urban development. However, when the economic benefits failed to materialize, city leaders proposed a mixed-use development with offices, restaurants, and residential property within the boundaries of the World Heritage Site, demonstrating the conflicting interests that stakeholders have within the framework of sustainable development (Labadi, 2017).

However, despite the general consensus within the research literature that short-term economic gain impedes the ability to develop sustainable tourism, we cannot simply focus on the negative impact that tourism brings to heritage resources (Helmy & Cooper, 2002). Instead, the conversation around sustainable tourism and heritage preservation must focus upon common goals of archaeologists, local communities, tourism developers, tourists and government agencies (Walker & Carr, 2013). From an archaeological perspective, because local communities are often left out of this conversation, archaeologists need to be proactive in their inclusion. They should be open to allowing local people to drive the research questions they believe would have applications to small-scale tourism, to be willing to provide full access to the results of their research, and to fund and assist in the creation of multilingual, tourist-friendly information that local guides can use, such as posters, pamphlets and guide maps (Glover *et al.*, 2012; Rissolo & Mathews, 2006), or 3D printed artifact replicas (McKillop & Sills, 2013). On a more ambitious level, archaeologists may provide support and resources for the creation of community-based museums (Ardren, 2002; Moser *et al.*, 2002).

Traditionally, most major archaeological sites and museums are run by government entities such as the National Park Service in the United States. During economic downturns, this means that austerity measures are put into place, and resources are distributed away from protecting and curating archaeological resources (Gould, 2017). This has also traditionally been a 'top-down professional approach' that generally ignores the views of stakeholders, including local communities and tourists in the process (Alazaizeh *et al.*, 2016b; Díaz-Andreu, 2013). More recently, there has been a push toward non-governmental, non-profit heritage management that embraces a more inclusive framework that provides benefits to local communities. Examples have consisted of non-profit organizational funding for site preservation and volunteer participation in archaeological research (Gould, 2017).

Community Engagement in Heritage

In 2003, UNESCO (United Nations Educational, Scientific and Cultural Organization) added a convention relating to intangible heritage and sustainable development. This was an attempt to shift the focus of heritage tourism to incorporate local communities and living culture into the equation. However, the criteria still concentrate overwhelmingly on preservation rather than development, and emphasize aspects of landscape design, architecture, town planning, settlements and technology. There is also a lack of inclusion of indigenous peoples in the process of decision making, often making it difficult for them to have the ability to engage in economic, cultural and social development as it relates to archaeological heritage (Castañeda & Mathews, 2013; Cayron, 2017; Gould, 2017; Hassan, 2017).

Instead, tourism development should focus on the protection of archaeological and environmental resources, while also benefitting local communities. Most of the indigenous communities living adjacent to cultural heritage sites are dependent upon the local environment for survival, including farming, fishing and the collection of forest resources. The communities usually live in poverty and place extreme pressure on the soils, water and forest resources for survival, further threatening heritage sites. It is therefore important to include development that allows the supply of key resources, such as clean water, within the broader tourism development plan (Wager, 1995). Locals should also not be expected to have to put the conservation of archaeological sites ahead of their own needs, including economic development (Holtorf & Ortman, 2008). Thus, if land use is going to be restricted for conservation purposes, other sources for producing food or natural resources must be put into place (Wager, 1995). Local peoples should also be given priority in heritage employment opportunities, and be encouraged to participate in the process of heritage protection that develops a healthy local economy, while protecting cultural assets (Jaafar *et al.*, 2015; Wager, 1995).

Unfortunately, thus far, sustainable tourism that benefits indigenous peoples has faced many challenges and pitfalls. First, local communities are vulnerable as the development efforts often fall short in terms of generating the jobs originally promised, or the jobs are limited to low-wage service positions that are usually seasonal. In places such as the Maya Riviera of Mexico, investors in tourism siphon off most of the profits, leaving little to trickle down into local communities (Díaz-Andreu, 2013; Pacifico & Vogel, 2012; Pyburn, 2017; Walker, 2009). Second, governments have to make difficult decisions as regards which development program to support and thus smaller heritage sites may not be a priority (Pyburn, 2017).

Similarly, government support for development efforts often shifts during economic downturns, which can even impact larger sites (Gould,

2017). Third, government tourism programs rarely directly involve local residents, and their engagement with tourists may be limited to encounters in restaurants or stores (Cayron, 2017). When programs do incorporate local people, determining who will represent the community can be challenging, as influential individuals or groups may dominate the decision-making, or governmental organizations may favor particular individuals who best suit their interests (Hassan, 2017). Fourth, many archaeologists who are interested in helping with tourism development in heritage places are only seasonally present in communities and have difficulty maintaining permanency in terms of providing permanent infrastructure, employment and support (Bawaya, 2005; Glover et al., 2012; Pyburn, 2017). Further, there is also less incentive in academia, particularly for junior scholars, to engage in community involvement, as it is not recognized, except in some instances, as something that counts towards tenure or promotion and takes away from the time and resources they could commit toward research and publication (Bawaya, 2005; Glover et al., 2012).

So, how do various stakeholders overcome some of these obstacles of developing community-based cultural heritage? First, we must understand that not all heritage sites should be developed for tourism. The reality is that some communities will not be interested in selling their past for tourism, or internal conflicts may result in the failure of a development project (Díaz-Andreu, 2013). However, consultation with communities is of utmost significance. For communities that do want to engage in heritage tourism, sustainable development that takes the opinions and beliefs of local communities into account has a better chance of long-term success in balancing the competing interests of environmental, economic and social needs (Jaafar et al., 2015; Pacifico & Vogel, 2012).

There is also a push that heritage resources should be recognized as common property of the community, allowing for more bottom-up organization that allows communities to avoid predation by corporations and governments. As archaeological resources have deep roots in communities, their intangibility should provide the opportunity to own and control them. In places like Italy and Spain, local communities face extreme challenges in gaining control over local heritage sites from the national and religious bodies that maintain them (Gould & Paterlini, 2017; González et al., 2017). However, these agencies might benefit from recognizing that the uniqueness of a cultural region is part of the draw for tourists. It can be a great advantage when the local community lives in harmony with the region's resources, is well versed in the local history, and views itself and its belonging to the locality with a sense of pride (Jaafar et al., 2015; Pacifico & Vogel, 2012). Traditionally, the representation of the indigenous past is often idealized and romanticized, and fails to appreciate contemporary indigenous peoples (Díaz-Andreu, 2013). Thus, modern communities interested in heritage tourism should have some say in the way in which cultural heritage is maintained, be active participants that interact with visitors, and present their culture from their perspective. These kinds of positive interactions can go a long way toward tourists appreciating contemporary indigenous culture and seeing this as an added value of a region (Cayron, 2017; Johnston, 2003). Unfortunately, in some locations (such as Africa) where heritage sites that have cultural and spiritual significance to local communities, heritage managers have criticized them for engaging in traditional ritual practices that have resulted in damage to the sites. In these cases, a less traditional model of custodianship may be needed, which incorporates indigenous knowledge systems and intangible cultural heritage. 'The objective of a traditional management system is generally to promote the sustainable use of both cultural and natural resources, and, by the same token, safeguarding the qualities and values of the site' (Jopela, 2011: 105).

Tourism development must also recognize that tourists may not even know that a cultural heritage site exists. This means that developers need to make an effort to raise awareness and demonstrate why it is a desirable place to visit. This endeavor should include reviewing the historical archive that documents all work that has been done previously at the site, including excavation, reconstruction, and conservation (Cleere, 2010). Once the information is gathered, it should be compiled into an easily digestible history that can be distributed via marketing strategies that will reach the target audience. This can include updated information in guidebooks and well-placed travel articles that lay out the amenities and points of local interest, social media and websites, and inclusion on local tour itineraries. Location near other tourism attractions is also a key in getting tourists to consider a visit (Cayron, 2017; Pyburn, 2017).

Additionally, a percentage of heritage site and museum entrance fees should be designated to assist local indigenous populations. For example, in China, portions of admission prices to sites such as Mount Wutai National Park and World Heritage Site are used to assist local peoples who have been displaced. In other regions that house sections of the Great Wall, funds are given to the local government for the maintenance of local communities. While this seems beneficial in theory, some have criticized that the high ticket prices preclude lower-income people from visiting their own cultural sites (Shepherd & Yu, 2013). This may necessitate designated days for free or discounted admission for local peoples, particularly school children and indigenous populations. In considering these issues of inclusion in heritage tourism development and management, it further forces us to wrestle with the realm of ethics.

Ethics and Archaeology-based Tourism

Traditionally, ethics in archaeology has focused on problems related to the documentation and conservation of heritage. However, within the broader realm of heritage tourism, this has extended to issues related to social justice, representation, authenticity and public engagement. Within the context of the Civil Rights movement in the United States, indigenous movements in places like Latin America and the rise of legislation such as the Native American Graves Protection and Repatriation Act (NAGPRA), archaeologists began to realize that there are variable ideas of how cultural patrimony is perceived and treated (Díaz-Andreu, 2013). In thinking about the many competing stakeholders that have interests in archaeological heritage, it is important to ask what the significance is to each and what their agendas are. For example, are they seen as valuable because of the economic returns they can bring, their outstanding representations of architecture or art that makes them potentially more marketable for tourism, or the valor they represent for the rich and powerful of the archaeological past (Hassan, 2017)? Or do they have value because of the educational opportunities they provide or the role they play in promoting national identity?

The archaeologist's agenda has been focused on documentation and preservation, and thus we have traditionally seen heritage tourism as a threat (Walker & Carr, 2013). Archaeologists have also been suspicious of the tourism industry, as the underlying profit motive is often going to be at odds with the central ethic of stewardship that pervades archaeology (Pacifico & Vogel, 2012). In other words, archaeologists are anthropologists with a relativistic heritage, meaning that the archaeological record is part of the public trust, and should be to the benefit of all people and for all time (Salmon, 1997). And yet, even this stewardship ethic within archaeology has changed in recent decades. As we have shifted from processual archaeology to post-processual archaeology, so has the way in which we view research design, ownership of the archaeological record and the information it yields. As Holtorf and Ortman (2008: 82) have argued, the use of resources toward the preservation of archaeological remains for future generations over urgent present-day needs is ethically debatable. 'The notion that we should save everything for the future when techniques will be better is a transparent absurdity, since the future, by definition, can never come... to impose this policy... is to empower state officialdom at the expense of the people to whom the heritage truly belongs.'

This post-processual approach also gives descendant populations equal standing with archaeologists in the production of knowledge creation. This is an approach that does not necessarily 'lock in' the meaning of an archaeological site, as definitions may change with new archaeological interpretations or perceived needs of contemporary populations (Walker & Carr, 2013). While professional ethical codes reflect these changes, they are made vague enough for archaeologists to flexibly interpret and implement their own ethical standards in terms of whether they rank the archaeological heritage over living people. As Pacifico and Vogel (2012: 1594–1595) state, 'Specifically a conflict between the priorities of conservation and other social justice values can arise depending on one's position on the meaning of 'stewardship,' conservation, and the belief in archaeology's potential to be universally beneficial'.

The ethical codes for the Society for American Archaeology (SAA) and American Anthropological Association (AAA) are set up to direct archaeologists in their relationship with heritage tourism. Although committed to researching, documenting and protecting archaeological resources, these codes make it clear that archaeologists do not control access to the archaeological record. There is a proposal that archaeologists should promote a mutually beneficial relationship with populations affected by archaeological research, and be receptive to other uses of the archaeological record, even when it clashes with archaeological interests (Pacifico & Vogel, 2012). One way in which archaeology has attempted to address these clashes and recognize the responsibilities that we have to descendant communities is through public archaeology. Public interest in archaeology is imperative to the field's survival, particularly as public funds are diminished. It is part of archaeologists' professional responsibility to explain to the general public the benefits of archaeology and heritage preservation (Sabloff, 1998; Walker & Carr, 2013), and to engage in public archaeology (Comer, 2012). Further, in the context of tourism, the public archaeology approach provides a stern warning not to develop sites if they cannot be properly protected. However, in developing countries this universal defending of a greater public good cannot always be justified. Instead, community-based archaeology asserts that community control at every step of development is a key ethical goal, emphasizing that archaeologists do not have exclusive control over archaeological resources (Pacifico & Vogel, 2012).

Finally, another ethical issue that has arisen in relation to tourism's use of heritage resources is the aspect of authenticity. In some cases, archaeological tourism development has been accused of 'Disneyfying' archaeology, in which cultural heritage no longer involves direct contact with the actual archaeological site or monument. For example, the original sites of the Cave of Altamira, Spain, and Lascaux Caves, France, have been closed down and replicas have been built nearby. This means that visitors are experiencing a 'new' cave, but in the same geographical and cultural context as the one that existed for tens of thousands of years. The Luxor Hotel in Las Vegas also houses a reproduction of Tutankhamun's tomb. Both of these examples offer visitors the sensation of being in an ancient archaeological space (Melotti, 2011). While these replicas help to protect the original site, some might argue that this is the ultimate form of elitism, that reserves the original for only the very few who are worthy of visiting it. However, others might argue that in today's society where virtual reality has become the norm, authenticity is relative (Melotti, 2011). It is certainly one extreme answer to the threat that heritage tourism brings to archaeological sites.

Balancing Tourism, Community Needs and the Protection of Sites

It is undeniable that visitors can cause irreversible damage to archaeological sites, forcing heritage managers to find ways to mitigate this harm. 'Direct management actions' address visitor behaviors and attempt to control those behaviors (Alazaizeh *et al.*, 2016b). In some cases, site managers have had to impose fines or sanctions, limit access to the site for tourists and local peoples by putting up fencing or eliminating the ability to touch the architecture, murals, rock art or other high-risk items (Alazaizeh *et al.*, 2016b; Soon, 2017), or limiting the number of entrance tickets (Alazaizeh *et al.*, 2016b). In the case of particularly fragile, dangerous, or sensitive cultural heritage sites this may be the only way to protect them (Alazaizeh *et al.*, 2016b). These actions, however, may have the result that tourists cannot see or experience what they came to see. Thus, they may lessen these restrictions somewhat by adding didactic panels or photographs with close-up images of murals, cave paintings or other delicate features (Soon, 2017).

Nonetheless, this direct approach has primarily emphasized how to control and limit tourist behavior, rather than examining tourists' motivations. More recent scholarship has moved toward 'indirect management approaches' that may inform visitors about appropriate behaviors on site, and attempt to change the decision-making factors on which visitors based their behaviors. Ensuring good visitor behavior is partly about meeting visitor expectations, while also raising awareness and sensitivity about the risk to fragile built environments (Duval & Smith, 2014). For example, rather than limiting the number of visitors who can enter a site, they may instead inform visitors about areas of the site that have potential crowding issues and suggest visiting alternative areas (Alazaizeh *et al.*, 2016b; Timothy & Boyd, 2003). Tarawneh and Wray (2017) suggest a tourism diversification strategy for Petra that includes the addition of Neolithic-period villages in the region. Including these smaller sites onto itineraries that target more adventurous tourists may increase the length of a visitor's stay, extend tourism dollars to communities outside of Petra, as well as provide conservation dollars to help preserve unique sites.

Some scholars have recommended that one way of improving indirect management approaches is to understand better who the tourists are and what their motivations are for visiting (Alazaizeh *et al.*, 2016b; Duval & Smith, 2014; Shepherd & Yu, 2013). First, they might consider whether visitors are primarily foreign or local/regional tourists. This can vary dramatically between regions and might impact the way in which heritage information would be presented. For example, according to the China National Tourism Administration, as much as 96% of tourists at sites like Huangshan are domestic (Shepherd & Yu, 2013: 61), whereas South African Tourism estimates that 30% of tourists in South Africa are foreign (Duval & Smith, 2014: 39). Contrarily, a random sampling of visitors to the site of Petra, Jordan, indicated that 91% were non-Jordanians (Alazaizeh *et al.*, 2016b).

Duval and Smith (2014) surveyed visitors to an area surrounding several rock art sites in South Africa and were able to identify five main types of visitors. Although their categories are specific to South African rock sites, they seem to be general enough to be applied to sites in general: (1) hedonists, who are drawn to family-friendly atmospheres for relaxation who want to engage in recreational activities with close proximity and ease of access; (2) outdoor and sports tourists, who attempt to get away from city life and are more interested in outdoor and cultural activities; (3) information seekers, whose motivations are to discover a new place, to be outdoors and somewhere different from where they live, and are driven by recommendations from tourist guides, websites and word of mouth; (4) cultural heritage enthusiasts who come specifically to visit cultural sites as a way of learning about history and different cultures; and (5) general sightseers, who are mostly international tourists motivated by discovering new things and getting a feel for the local culture.

Additionally, they argue that there are several ways to promote cultural tourism that will ultimately help promote and protect heritage. First, guided tours with accredited (perhaps indigenous) guides can instruct visitors about site etiquette while teaching them about the significance of a cultural site. It is particularly important to incorporate local peoples into tourism development to encourage indigenous stewardship and maintain or reinvigorate the social value of the sites, particularly where religious or cultural significance may have been lost (Duval & Smith, 2014). Cultural heritage sites can also be promoted for their unique qualities or in a comparative manner that would allow visitors to understand how they relate to other sites in the region, which would help visitors understand the region better. Tourism promoters can also highlight the accessibility of cultural heritage sites and promote it to the appropriate audience, such as selling easily-accessible and family friendly sites to 'hedonist' tourists and less accessible sites to outdoor and sports tourists.

Conclusion

In conclusion, this chapter has attempted to emphasize some of the pitfalls, lessons learned and best practices of archaeologists and tourism specialists from around the world. In recent decades, archaeologists have become considerably more active in the process of developing and promoting sites for heritage tourism. In some cases, they do so as a way to protect the past in the face of uncontrolled tourism. In other cases, they may view it as an essential aspect of human rights and social justice concerns (Díaz-Andreu, 2013; Pyburn, 2017). Regardless of their motives, as we move forward in the field of cultural heritage tourism, scholars should produce research that can be broadly applied, and that is not simply focused on isolated regions with esoteric issues so that we do not continue to reinvent the wheel (Loulanski & Loulanski, 2011).

References

Alazaizeh, M.M., Hallo, J.C., Backman, S.J., Norman, W.C. and Vogel, M.A. (2016a) Crowding standards at Petra Archaeological Park: A comparative study of McKercher's five types of heritage tourists. *Journal of Heritage Tourism* 11(4), 364–381.

Alazaizeh, M.M., Hallo, J.C., Backman, S.J., Norman, W.C. and Vogel, M.A. (2016b) Value orientations and heritage tourism management at Petra Archaeological Park, Jordan. *Tourism Management* 57, 149–158.

Ardren, T. (2002) Conversations about the production of archaeological knowledge and community museums at Chunchucmil and Kochol, Yucatán, México. *Journal of World Archaeology* 34 (2), 379–400.

Bawaya, M. (2005) Maya archaeologists turn to the living to help save the dead. *Science* 309 (5739), 1317–1318.

Castañeda, Q.E. and Mathews, J.P. (2013) Archaeology Meccas of tourism: Exploration, protection, and exploitation. In C. Walker and N. Carr (eds) *Tourism and Archaeology: Sustainable Meeting Grounds* (pp. 37–64). Walnut Creek, CA: Left Coast Press.

Cayron, J.C. (2017) Archaeological heritage tourism in the Philippines: Challenges and prospects. In P.G. Gould and K.A. Pyburn (eds) *Collision or Collaboration: Archaeology Encounters Economic Development* (pp. 89–102). Cham, Switzerland: Springer.

Cleere, H. (2010) Management plans for archaeological sites: A World Heritage template. *Conservation and Management of Archaeological Sites* 12 (1), 4–12.

Comer, D.C. (2012) *Tourism and Archaeological Heritage Management at Petra: Driver to Development or Destruction?* Dordrecht: Springer.

Díaz-Andreu, M. (2013) Ethics and archaeological tourism in Latin America. *International Journal of Historical Archaeology* 17 (2), 225–244.

du Cros, H. and McKercher, B. (2015) *Cultural Tourism* (2nd edn). London: Routledge.

Duval, M. and Smith, B.W. (2014) Seeking sustainable rock art tourism: The example of the Maloti-Drakensberg Park World Heritage Site. *The South African Archaeological Bulletin* 69 (199), 34–48.

Glover, J.B., Rissolo, D., Mathews, J.P. and Furman, C.A. (2012) El Proyecto Costa Escondida: Arqueologia y Compromiso Comunitario a lo Largo de la Costa Norte de Quintana Roo. (The Costa Escondida Project: Archaeology and community engagement along Quintana Roo's north coast, Mexico) *Chungara, Revista de Antropología Chilena* 44 (3), 511–522.

González, P.A., Macías Vázquez, A. and Fernández, J. (2017) Governance structures for the heritage commons: La Ponte-Ecomuséu-Ecomuseum of Santo Adriano, Spain. In P.G. Gould and K.A. Pyburn (eds) *Collision or Collaboration: Archaeology Encounters Economic Development* (pp. 153–170). Cham, Switzerland: Springer.

Gould, P.G. (2017) Collison or collaboration? Archaeology encounters economic development: An introduction. In P.G. Gould and K.A. Pyburn (eds) *Collision or Collaboration: Archaeology Encounters Economic Development* (pp. 1–13). Cham, Switzerland: Springer.

Gould, P.G. and Paterlini, A. (2017) Governing community-based heritage tourism clusters: I Parchi della Val di Cornia, Tuscany. In P.G. Gould and K.A. Pyburn (eds) *Collision or Collaboration: Archaeology Encounters Economic* Development (pp. 137–152). Cham, Switzerland: Springer.

Hassan, F.A. (2017) The future of heritage management: Ethics and development. In P.G. Gould and K.A. Pyburn (eds) *Collision or Collaboration: Archaeology Encounters Economic Development* (pp. 14–27). Cham, Switzerland: Springer.

Helmy, E. and Cooper, C. (2002) An assessment of sustainable tourism planning for the archaeological heritage: The case of Egypt. *Journal of Sustainable Tourism* 10 (6), 514–535.

Holtorf, C. and Ortman, O. (2008) Endangerment and conservation ethos in natural and cultural heritage: The case of zoos and archaeological sites. *International Journal of Heritage Studies* 14 (1), 74–90.

Howell, G. (2006) *Gertrude Bell: Queen of the Desert, Shaper of Nations*. New York: Farar, Strauss and Giroux.

Jaafar, M., Noor, S.M. and Rasoolimanesh, S.M. (2015) Perception of young local residents toward sustainable conservation programmes: A case study of the Lenggong World Cultural Heritage Site. *Tourism Management* 48, 154–163.

Johnston, A.M. (2003) Self-determination: Exercising indigenous rights in tourism. In S. Singh, D.J. Timothy and R.K. Dowling (eds) *Tourism in Destination Communities* (pp. 115–134). Wallingford: CABI.

Jopela, A. (2011) Traditional custodianship: A useful framework for heritage management in southern Africa? *Conservation and Management of Archaeological Sites* 13 (2–3), 103–122.

Labadi, S. (2017) UNESCO, World Heritage, and sustainable development: International discourses and local impacts. In P.G. Gould and K.A. Pyburn (eds) *Collision or Collaboration: Archaeology Encounters Economic Development* (pp. 45–60). Cham, Switzerland: Springer.

Loulanski, T. and Loulanski, V. (2011) The sustainable integration of cultural heritage and tourism: A meta-study. *Journal of Sustainable Tourism* 19 (7), 837–862.

Mairs, R. and Muratov, M. (2015) *Archaeologists, Tourists, Interpreters: Exploring Egypt and the Near East in the Late 19th-Early 20th Centuries*. New York: Bloomsbury.

McKercher, B. (2002) Towards a classification of cultural tourists. *International Journal of Tourism Research* 4 (1), 29–38.

McKillop, H. and Sills, E.C. (2013) Sustainable archeological tourism of the Underwater Maya Project by 3D Technology. *Anthropology News* May/June, 12–13.

Melotti, M. (2011) *The Plastic Venuses: Archaeological Tourism in Post-modern Society*. Newcastle: Cambridge Scholars Publishing.

Moser, S., Glazier, D., Phillips, J.E., Nasser el Nemr, L., Aaleh Mousa, M., Nasr Aiesh, R., Richardson, S., Conner, A. and Seymour, M. (2002) Transforming archaeology through practice: Strategies for collaborative archaeology and the community archaeology project at Quseir, Egypt. *World Archaeology* 34 (2), 220–248.

Orbaşli, A. and Woodward, S. (2012) Tourism and heritage conservation. In T. Jamal and M. Robinson (eds) *The SAGE Handbook of Tourism Studies* (pp. 314–332). London: Sage.

Pacifico, D. and Vogel, M. (2012) Archaeological sites, modern communities, and tourism. *Annals of Tourism Research* 39 (3), 1588–1611.

Pyburn, K.A. (2017) Developing archaeology. In P.G. Gould and K.A. Pyburn (eds) *Collision or Collaboration: Archaeology Encounters Economic Development* (pp. 189–199). Cham, Switzerland: Springer.

Rissolo, D. and Mathews, J.P. (2006) Archaeologists working with the contemporary Yucatec Maya. In J.P. Mathews and B. Morrison (eds) *Lifeways in the Northern Maya Lowlands: New Approaches to Archaeology in the Yucatán Peninsula* (pp. 198–209). Tucson: University of Arizona Press.

Sabloff, J.A. (1998) Distinguished lecture in archaeology: Communication and the future of American archaeology. *American Anthropologist* 100 (4), 869–875.

Salmon, M.H. (1997) Ethical considerations in anthropology and archaeology, or relativism and justice for all. *Journal of Anthropological Research* 53 (1), 47–63.

Shepherd, R.J. and Yu, L. (2013) *Heritage Management, Tourism, and Governance in China: Managing the Past to Serve the Present.* New York: Springer.

Soon, S.C.M. (2017) Protection and conservation of archaeological heritage in Malaysia: Issues and challenges. In P.G. Gould and K.A. Pyburn (eds) *Collision or Collaboration: Archaeology Encounters Economic Development* (pp. 29–43). Cham, Switzerland: Springer.

Stephens, J.L. (1969) *Incidents of Travel in Central America, Chiapas and Yucatan, Volumes I and II.* Mineola, NY: Dover Publications.

Tarawneh, M.B. and Wray, M. (2017) Incorporating Neolithic villages at Petra, Jordan: An integrated approach to sustainable tourism. *Journal of Heritage Tourism* 12 (2), 155–171.

Timothy, D.J. and Boyd, S.W. (2003) *Heritage Tourism.* London: Prentice Hall.

Wager, J. (1995) Developing a strategy for the Angkor World Heritage Site. *Tourism Management* 16 (7), 515–523.

Walker, C. (2009) *Heritage or Heresy: Archaeology and Culture on the Maya Riviera.* Tuscaloosa: The University of Alabama Press.

Walker, C. and Carr, N. (eds) (2013) *Tourism and Archaeology: Sustainable Meeting Grounds.* Walnut Creek, CA: Left Coast Press.

11 Interpreting the Past: Telling the Archaeological Story to Visitors

Sue Hodges

Introduction

Archaeology is undoubtedly the sexiest of the heritage professions, conjuring up Indiana Jones, Tutankhamun and Pompeii. However, the idea of being a detective of the past who uncovers buried treasures, fights tomb raiders and finds a way to break curses is far removed from the prescriptive rules for archaeological interpretation that have often prevailed in the heritage sector.

The story of how archaeological ruins have been interpreted is the story of a growing realisation among heritage professionals that the past does not simply 'speak' for itself but is mediated and determined by who is telling the story. This change has been accompanied by a tension between the desire to preserve archaeological relics, on the one hand, and the desire to display, educate, entertain and inform through interpretation, on the other. Because there is little specific literature on archaeological interpretation, this chapter draws on an analysis of trends in the heritage sector and the field of heritage interpretation to frame my understanding of how interpretation has worked in the field of archaeology.

Changing Philosophies of Interpretation

It is axiomatic in the historical discipline that the past is political, and that history is written by the victor, but an understanding of this has only emerged recently in the field of archaeology. Archaeological relics were initially displayed in the service of the nation-state. Manipulations of archaeology for both Slavic and German nationalistic purposes began as early as the 16th century and were matched in England by the study of British antiquities in 1533 and by the display of archaeological relics at the British Museum in 1753 (Fowler, 1987). As European travel became common with the Grand Tour in the 17th and 18th centuries, archaeological ruins became the vehicle for one-upmanship between countries. They were also the nexus for political machinations; under Mussolini's dictatorship in the 20th century, for instance, Italy's fascists interpreted Pompeii's ruins as evidence of Italy's past and future glory (Downs *et al.*, 2009).

Before the 20th century, archaeology was often a pastime of the elite. During the 1860s, Pompeii's remains became a showcase for the new kingdom of Italy and were accessible only to the rich until archaeologist Giuseppe Fiorelli opened up the site for everyone. (Downs *et al.*, 2009). In 19th-century England, romantic poets such as Wordsworth, Keats, Shelley and Byron encouraged their readers to 'ramble' among ruins such as Tintern Abbey in England and the Acropolis in Greece. Many of these poets had classical educations and their audiences had no need for interpretation, since they brought with them on their travels a knowledge of history, literature, the classics and antiquity.

Romanticism also formed the basis of the early conservation movement, which evolved to protect material fabric from change (Araoz, 2011). In his 1877 *Manifesto of the Society for the Protection of Ancient Buildings*, William Morris wrote that 'restoration' would lead to the 'appearance of antiquity' being taken away from old parts of a ruined building that were left, resulting in a 'feeble and lifeless forgery' (SPAB, 2019). Almost 90 years later, the 1964 ICOMOS *Venice Charter* was similarly based on the notions that the value of places lay entirely in their material fabric (Young, 2013) and that data was objective and could be 'read' statistically to provide truth about the past (Schorch, 2017). Within the discipline of archaeology, this translated into processual, behavioural, ecological, evolutionary and positivist views dominating theory and practice and leading to limited, scientifically based interpretations (Schorch, 2017). Within this schema, traditional means of recording archaeological data were often understood to be unproblematic 'archaeological reality', despite the fact that image-making (much like the process of archaeology itself) is inherently subjective and creative and that carefully framed photographs of sites routinely excluded the people and tools linked to the excavation (Watterson, 2015).

The profession of heritage interpretation, which began in the United States, appears to have had little influence on archaeological interpretation but had similar romantic origins to the historic preservation movement. In the 1920s, naturalist John Muir (1912) used the word 'interpret' for the first time in association with nature:

> I'll interpret the rocks, learn more about the language of flood, storm and the avalanche. I'll acquaint myself with the glaciers and wild gardens, and get as near the heart of the world as I can. ((Muir, 1912), cited in Widner Ward & Wilkinson, 2006: 7)

The interpretation profession grew strongly in the US parks sector with the aim of conserving natural places and fostering public stewardship of cultural resources (Jameson, 2007). In 1957, US National Park Service staff member Freeman Tilden wrote his seminal *Interpreting Our Heritage*, which still remains the 'Bible' for many heritage interpretation professionals today. Tilden's key principles – that the chief aim of interpretation is provocation and that interpretation is an art encompassing 'revelation based on information' (Tilden, 1957) – situate interpretation within an essentialist framework: interpreters communicate key messages about a place based on 'raw material' and 'facts' provided by historians, archaeologists and architects (Silberman, 2013). Interpreters influenced by Tilden's views have subsequently developed models for interpretive communication based on messages, themes and stories to be delivered at sites such as museums, aquaria, national parks and zoos, where the role of interpretation is to educate and inform. Frequently this interpretation is communicated by trained visitor guides, through education programs, through displays at visitor centres and along signage trails. It is often accompanied by a focus on visitor research, communication techniques and cognitive psychology. However, archaeological sites and other forms of cultural heritage fabric have sometimes been seen as antithetical to the idea that places are 'wild' and untouched.

The Tilden-led approaches outlined above only partially meet the needs of archaeological interpretation in the 21st century, which increasingly takes place on large, complex sites and is being called upon to address issues such as refugees, historical revisionism, climate change, job creation and place making. As John Jameson (1997: 86) has stated, 'archaeology has an ethical imperative to make the past accessible to the public and to empower people to participate in a critical evaluation of the pasts that are presented to them'. Moreover, both romantic essentialism and a scientific, processual and positivist approach deny the fact that archaeological relics are only one manifestation of a past social structure and that intangible elements of that culture – rituals, rites and cultural processes – produced them (Hodder, 1991).

So, how can these long-lost traces of past lives be captured without straying too far into the realms of fiction? Good archaeological interpretation begins by encouraging visitors to share in the excitement of discovering relics and ruins and to imagine the lives of the people who once lived there. For Kiri Sharpe, a trained archaeologist with an understanding of Maori spirituality who excavated in Āotearora New Zealand, going on a site visit was always exciting because people from the past had been 'just where you are standing' (quoted in Pishief, 2017: 59). Good heritage interpretation also considers 'messy' social histories as central to its cause (Gonzales-Tennant, 2011). These histories can have complex, content-rich narratives; disputed claims for authority; multiple and conflicting facts and be told from many points of view, all of which help visitors realise that history is a living, ever-changing and subjective discipline.

This understanding was behind Grace Karskens' pioneering work on archaeological interpretation in the early 1990s. She used a combination of ethnographic and social history research to unravel the private lives of 19th-century residents of The Rocks, an inner Sydney foreshore area. This is a place with many histories: first as the home of the Darug people; next as the site of the first European settlement in Sydney; later as its commercial heart; still later as a slum, and today as one of Sydney's main tourist areas. Karskens (1999) focused on how archaeological remains of food, housing, bodily functions and care all revealed how people lived in the past. As a case study, she investigated how rubbish found in a well behind the house of 19th-century bigamist George Cribbs provided clues to his turbulent domestic life. Because the well contained some relatively expensive broken tumblers, Karskens theorised that Cribbs's second wife, Fanny, might have smashed the China on impulse when she heard that his first wife, Mary, to whom he was still married, had set sail from England to Australia. Karskens based her theory on the fact that a Sydney mender of china, Constable John Justice, was well known for his talent of 'alleviating the otherwise fatal and irreparable effects of domestic inquietitude…' (Karskens, 1999: 138). While this insight into domestic life in early Sydney is speculative, it does show how the 'immediacy of the physical, tactile dimension' connects visitors to people from the past who perhaps had the same feelings as we do today (Karskens, 1999: 17).

Changing Meanings

Two of the major influences on interpretation over the last few decades have been that the archaeologist or historian is always a part of the interpretive story and that interpretation is not a scientific process divorced from wider meanings and the people who inhabited a place (Hodder, 1991; Pishief, 2017). In the late 1980s and early 1990s, ethnographers and archaeologists began introducing 'I' into their work in an attempt to understand their relationship with archaeological material. The process of archaeology also began to be seen as discursive and recursive by practitioners: a performance where the process of archaeological excavation itself affected interpretation (Cobb & Croucher, 2014). In turn, this self-reflexivity enabled archaeologists to take part in conversations about how to interpret the past with the communities and/or Indigenous groups directly associated with the heritage sites in question (Hodder, 1991).

From the 1970s onwards, archaeologists also began to focus not only on the internal tensions and conflicts that produced social change in a community but also on the roles played by individuals in these dynamics (Saitta, 1994). This 'new archaeology' questioned *what* was being interpreted and was concerned with the epistemological grounds and content of interpretation (Ablett & Dyer, 2009). Rather than perceiving the landscape as a series of isolated monuments, these archaeologists also interpreted sites as cultural landscapes that expressed intricate patterns of social power, ideology and gender (Saitta, 1994: 203). This change in perspective occurred at the same time as the New Social History movement emerged in the field of history. This movement, which continues today, challenged the dominant narratives of power, colonialism and patriarchy with counter-narratives of Indigenous people, women, working-class people and other previously marginalised groups, with the aim of giving voice and agency to people who had been left out of the historical record. In parallel with this, the idea of the public has also grown and changed. In the fields of archaeology and people, the term 'public' now includes local residents and groups, and people who were previously sidelined, oppressed or excluded from participating in interpreting their pasts, together with tourists and students (the traditional audiences for interpretation). Now, all of these different stakeholders have roles not only in creating meaning at archaeological sites but also in consuming them as customers (Gonzalez-Tennant, 2011). However, the term 'public archaeology', which was first used in Charles McGimsey's (1972) eponymous book, did not reflect these concerns but focused instead on stewardship of cultural resources (Gonzales-Tennant, 2011). This linked to the idea that the public needed to be educated about an archaeological site so that they could understand its (scientific) importance (Pishief, 2017).

Public interpretation was met with resistance by some archaeologists including Gillian Binks (1986), who explained that the chief duty of archaeologists was preservation rather than public interpretation (Robb, 1998). The tendency of heritage managers and consultants in the 1980s for 'bricolage' – a combination of pastiche, anachronism and eclecticism – also provoked much criticism (Robb, 1998). Nevertheless, the recognition that archaeology was funded by the public and therefore had a responsibility to communicate to that public sowed the seeds for modern ideas of the importance of interpretation and public programs relating to archaeology (Gonzales-Tennant, 2011).

At the same time, the heritage conservation movement, as embodied by International Council on Monuments and Sites (ICOMOS) charters and other official documents, showed the influence of these new theories. As early as 1903, Alois Riegl had explored how accepting different sets of values could lead to vastly different outcomes in conservation practice. However, the 1964 ICOMOS *Venice Charter* codified only two values as needed for heritage designation: historical and aesthetic (Araoz, 2011). Three decades later, ICOMOS heritage charters incorporated dramatically new ways of understanding the meanings and values of heritage sites. These included the concept of the inter-generational transmission of knowledge (ICOMOS *Charter on the Built Vernacular Heritage, 1999)*, the idea that reconstruction could be valuable for experimental research and interpretation (ICOMOS *Charter for the Protection and Management of the Archaeological Heritage, 1990)* and the idea that authenticity is culturally relative (ICOMOS *Nara Document on Authenticity, 1994)* (Araoz, 2011). The 2007 ICOMOS *Charter for the Interpretation and*

Presentation of Cultural Heritage Sites also contains guidelines on authenticity and inclusiveness.

But what does the public actually make of archaeological interpretation? In 2015, Roos Nagtegaal undertook a study to ascertain how local residents engaged with Roman heritage in the Netherlands, a country where Roman history is largely invisible and where interpretation may therefore take on a greater importance than in countries where archaeological remains are visible (Nagtegaal, 2015). His study found that some Nijmegen residents engaged passionately with their local archaeological heritage to the point where it became part of their sense of identity, and where they believed that the Roman ruins in their backyards 'belonged' to them (Nagtegaal, 2015). However, information panels in the town were elitist – the chair of the neighbourhood committee even stated that telling Roman histories would be '(casting) pearls before swine' – and therefore difficult for most people to understand (Nagtegaal, 2015: 50). By contrast, some older local residents reported that videos embedded in 'time gates' gave them a good view of Roman times and connected the history of their house to Roman history in general, while younger people preferred digital representations (Nagtegaal, 2015).

Multiple Meanings

> Communities use the past in myriad ways.... Quickly vanishing are the days when archaeologists believed their work could escape the politics of the modern world. Archaeological methods, data, and interpretations are now recognised as inherently political forms of knowledge, produced in relation to modern social and cultural constraints. (Gonzales-Tennant, 2015: 12)

This idea that the meanings of an archaeological site can be generated by local communities is also central to contemporary interpretation. The *Burra Charter 1999: the Australia ICOMOS Charter for Places of Cultural Significance* includes the idea that stakeholder communities have a role in interpreting and conserving places of special importance to them. Gustavo Araoz (2011) has described this as a 'new heritage paradigm' that officially recognises that heritage sites may have little or no material fabric to preserve, where social processes are integral to the significance of the place, where heritage places can be tools for poverty reduction and where facsimile reconstructions are accepted as 'valid equivalents' of originals long gone (Araoz, 2011: 55). His caution against the 'extreme anastylosis of archeological ruins justified as interpretation to make archeological sites more attractive and intellectually accessible' (Araoz, 2011: 56) captures the concern of many heritage professionals that interpretation will be used as a substitute for the 'real thing'. At the 2016 'Life Beyond Tourism' conference, for example, the then-ICOMOS Vice-President Peter Phillips cautioned against new ideas and warned that smart technologies could 'open the gates of hell'.

An alternative approach is to question who the archaeological remains are for – the archaeologist, the community, the scientific community, the visitor or all of the above – and tease out how interpretation can respond to the needs of each stakeholder. This may be as simple as changing the word 'site' to 'place'. As archaeologist and cultural theorist Laurajane Smith (2008) has commented, this will allow heritage to embrace multiple kinds of significance, assist in identity formation and 'anchor shared experiences' in ways that archaeological and other 'sites' do not (cited in Opp, 2011: 243).

Co-created interpretation, with its attendant self-reflexive methodology, also offers some solutions to capturing the voices and values of multiple stakeholders. It began with the realisation, in the late 1980s and early 1990s, that subordinate groups did not necessarily want to fit their interpretations into the scientific frameworks of Western cultural institutions because monuments and relics had personal meanings for them (Hodder, 1991). For instance, the Niitsitapi of southern Alberta, Canada, commented that the significance of the Writing-on-Stone archaeological site could not be determined by archaeologists:

> Áísínai'pi is a sacred place because spirits are present in the valley. The cultural and spiritual significance of Áísínai'pi does not come from the presence of archaeological evidence or specific physical components. It doesn't matter how many rock art or archaeological sites are located at Áísínai'pi, or how they are described and catalogued. The landscape and our traditions make Áísínai'pi significant to the Niitsitapi. (cited in Opp, 2011: 251)

However, archaeologists, in common with many heritage practitioners, have sometimes distanced themselves from these spiritual and emotional connections to place, the very elements that are not only the most important element of the site for Indigenous communities (Pishief, 2017) but may also have the most potency for visitors. Changing this dynamic means accepting that venerated scientific fabric may not be significant at all for other people. In the words of the Māori in New Zealand, when speaking of the difference between Pākehā (white people's) and Māori understandings of place:

> *Wāhi tapu* are the connections with the important landmarks, not the tangible place itself as Pākehā usually interpret it, but the connection between the place and people ... When we talk about *wāhi tapu* we are not just talking about the physical layout of the *pā*, we are not just talking about what you can see, we are talking about those relationships, we are talking about the spiritual side, we are talking about the events that happened here, so all those, all those concepts we see as the *wāhi tapu*. (Pishief, 2017: 66)

Storytelling by local communities has provided a powerful way of expressing these alternative meanings, as well as providing counter-narratives to 'official' histories (Gonzales-Tennant, 2011). Moreover, it has acted as a corrective to the Western-centric, scientific archaeological interpretation that ignored sources of evidence such as social history, ethnography and oral histories. James Opp, in his work on Writing-on-Stone, cites text from a c1935 sign that interpreted the adjacent archaeological remains:

> The petroglyphs or picture writings inscribed on the rocks in this vicinity are the works of an ancient race which inhabited this area prior to the advent of the white man. (quoted in Opp, 2011: 247)

In doing so, the sign perpetuated what Opp (2011) describes a 'colonial archeological gaze' and ignored recent evidence that linked the rock art to events in the 18th and 19th centuries. In the 1990s, steps by provincial officials to involve the local Aboriginal community in reinterpreting the site dramatically reshaped and remade Writing-on-Stone as a Niitsitapi place (Opp, 2011).

Nevertheless, the idea that oral records and legends are appropriate for sites has not met with universal approval. Stonehenge is a case in point. In 1984, *This is Spinal Tap*, a 'mockumentary' about a fake heavy metal band, featured a song *Stonehenge*. In one of the most famous scenes from the film, the band ponders the mystery of the Druids from a vastly under-scale replica of the famous site: 'No one knows who they were or what they were doing/But their legacy remains/ Hewn into the living rock, of Stonehenge' (Guest *et al.*, 1984). These lyrics encapsulate the many thousands of words written by heritage practitioners on the conflict between tangible and intangible heritage at the ancient site.

While archaeologists interpret the site scientifically, druids, pagans and priest-astronomers see it as mystical. Interpretations of Stonehenge as an 'astronomical observatory' built by scientifically advanced people and used by Indigenous astronomer-priests have attempted to bridge these vastly different worlds, even though there is little evidence for these activities (Fowler, 1987). Archaeologists Peter Stone and Robert MacKenzie (1989) opted for the middle ground in the 1990s, stating that both academic and mystical opinions should be included in the interpretation of Stonehenge:

> To do less, to simply display Stonehenge in 'romantic isolation' or only disseminate the contemporary scientific and archaeological understanding of the monument is to remove much of the relevance to today's society of arguably Britain's most famous landmark. (quoted in Robb, 1998: 591)

Some archaeologists take a dual role in an attempt to embrace the notions that archaeological interpretation can contain conflicting and irresolvable meanings and be co-curated. On the one hand, they are cultural actors involved in finding and interpreting material. On the other, they act as cultural intermediaries, information brokers and translators of meanings for associated communities (Schorch, 2017). In this model, communities are recognised as always divided, conflicted and multi-vocal, particularly in the case of local versus tourist communities (Tunbridge, 2017). They are also not geographically bound: there may be communities of archaeologists, just as there are village communities (Pishief, 2017). Outcomes from heritage interpretation, such as an increase in visitor numbers, job creation and social capacity-building, can also be embedded in planning from the outset and explicitly direct the interpretation undertaken on site.

If the need for different modes of interpretation to occur simultaneously is recognised, heritage interpretation can be a 'profoundly important public activity' as a public discourse (Silberman, 2013: 7) and serve both civic and citizen interests (Gonzales-Tennant, 2011). Importantly, sharing findings among those invested in the site – the archaeological community, local communities and Indigenous owners – can help heal rifts and empower subordinated groups (Gonzales-Tennant, 2011). Interpretation shared with communities has traditionally taken the form of illustrations, site photographs, presentations, talks, displays, pamphlets and discussions but now includes games, performances and three-dimensional (3D) virtual simulations of the site (Gonzales-Tennant, 2011). As Gonzales-Tennant (2011: 37) points out, interpretation is crucial because placing raw data online 'does little to motivate the general public to explore archaeological research'. Importantly, for archaeologists working with communities this also means a commitment to long-term sustained relationships that occur across the whole lifecycle of a project, from pre-planning stages through to 'interpretation and dissemination' (Gonzales-Tennant, 2011: 30–31, 34–35).

More radically, John Giblin (2015), in his work on post-conflict heritage after the Rwandan genocide, contends that archaeologists must also take responsibility for future uses of their work. He argues that archaeology has not been an innocent bystander in colonial and post-colonial constructions of the past and that to recourse to 'professional ethics' is an abnegation of the archaeologist's ethical responsibilities in interpretation. In his view, dynamic archaeological ethics must move beyond concern for 'the record' and instead consider the future uses of archaeology, including ones where new archaeologies produced may become implicated in future structural and physical violence (Giblin, 2015).

Interpretive Media

Archaeological interpretation is of course as much about the media used for interpretation as it is about the kind of stories told. In line with the rest of the heritage sector, 1980s archaeological interpretation was usually told offsite through television and print media (Hoffman *et al.*, 2002). Site-based interpretation featured a range of static media, such as panel interpretation, and sometimes visitor information centres that contained curated displays, plaques, panel signs, brochures, guidebooks, films and artworks. As museums and visitor centres developed in the 1990s, some exhibits contained large-scale reproductions of key events together with a film, push button lighting displays, special effects such as a Pepper's Ghost and often a 'book on a wall' as curators struggled with storytelling in a 3D space. If drafted in the absence of authorship or an awareness of the need to include multiple voices, interpretation was often didactic and based on 'facts'. In best practice cases, interpretation contested popular myths with evidence: the mid-1990s English Heritage display on Tintagel in Cornwall, UK, meticulously deconstructed the idea that Tintagel was Camelot and posed an alternative hypothesis that Arthur existed outside medieval romances (Robb, 1998). Twenty years later, the need to make money from legends was perhaps more important than evidence-based interpretation. In 2016, English Heritage controversially carved Merlin's face into the cliff below Tintagel Castle and its shop was dominated by Arthurian souvenirs.

Some early archaeological museums explored the notion of uncovering archaeological ruins on site, which has since become a common practice in archaeological interpretation. In the 2000s, Hyde Park Barracks in Sydney, Australia, placed transparent coverings over *in situ* remains of the archaeological investigation that had uncovered the foundations of the site and then explored what these remains meant through a series of lively, mixed-media displays. One remarkable feature of the interpretation – probably never repeated anywhere else in the world – were two rats in a cage in the ticketing area of the visitor centre. The role of the rats, 'Typhus' and 'Scurvy', was to encourage people to ask questions about the building. What was the connection between the rats and the history of Hyde Park Barracks? As it turned out, rats had scavenged clothing and other fabric for their nests in the 19th century, which is how we know now what the convicts and soldiers wore. Sadly, Typhus and Scurvy are no longer on display, possible victims of 21st-century Australian workplace health and safety regulations.

But perhaps Hyde Park Barracks' most famous display was the 'ghost stairway' – a wire frame that traces the outline of a ruined staircase. In this interpretation, site managers encouraged visitors to use their imaginations to reconstruct the missing parts of the staircase. This is now a popular form of archaeological interpretation. At Moerenburg in the city of Tilburg in the Netherlands, for example, 3D interpretive works in Corten – described as 'scenic embeddings' – trace the outlines of lost elements of the site. Visitors are invited to fill in the gaps by using information contained in the accompanying interpretive panels, which gives them an active part to play in assembling the story of the site and also makes them aware that sometimes only fragments of the past remain.

Because the use of digital technology in interpretation was also minimal until the late 1990s/early 2000s, little contestation of official narratives was possible except through the following formats: visitors' books, face-to-face tours where interpretation was refreshed as new evidence came to light, and interactive touchscreens. This was also the era of the growth in 'heritage centres' at 'compound sites': places where the focus was not just on one archaeological monument but on archaeological attractions embedded in a locality (Robb, 1998). In the UK, sites including Jorvik Viking Centre, Stonehenge and Tintagel set the scene for the battle between traditional and modern approaches to archaeological interpretation (Robb, 1998). Essentially, this was all about the value and role of archaeology. Should it be educational alone, or was entertainment sullying sites and creating 'Archaeological Disneylands' (Barry, 2014)? The dichotomy is evident in Kristen Barry's (2014: 42) comment that exciting new technological developments could make archaeological ruins less 'boring' but that this could pose a problem with sites intended to be educational. Statements like this not only show a limited understanding of educational theory but also indicate a view that authentic interpretation cannot be a popular endeavour. Nevertheless, Barry (2014: 46) does acknowledge the complexity of the issue by commenting that modern design and virtual reconstructions help create a 3D experience for visitors 'who may not otherwise identify easily with archaeological foundations'.

At the time of writing, tourism has proved a mixed blessing for archaeological sites. At the 2001 meeting of the Society for American Archaeology (SAA), the Public Education Committee identified tourism as a topic for further investigation because it offered a significant outreach opportunity for archaeology (Hoffman *et al.*, 2002). A year earlier, a SAA survey had found that archaeological parks were particularly important educational tools because they provided the only first-hand experience of an archaeological site for most people (Hoffman *et al.*, 2002). In the 1980s, interpretation also began taking a role as a corrective to the growing commercialisation of archaeological heritage, which some archaeologists controversially believed was due to the 'professional neglect' of public interpretation (Wickham-Jones, 1988). But this may now have become a poisoned chalice. In some developing countries, archaeological tourism has become a grab-bag for improving the economy of local communities. A 2003 USAid report on Jordan's tourism economy stated that 'Iconic heritage and landscapes have a unique role as keystone building blocks of quality visitor experiences and powerful motivators supporting tourism marketing success' (Comer & Willems, 2011: 515–516). Similarly, in 2014 the Asian Development Bank put out a large tender for the interpretation and presentation of cultural heritage sites in Punjab, India, with the intent of using interpretation of cultural and natural heritage sites to stimulate Punjab's tourism development. It is difficult, and unethical, for Western professionals to argue against this given their privileged economic positions, yet insensitive and highly-commercial interpretation has placed the future of sites such as Gobindgarh Fort in Punjab at risk, which will not serve anyone.

To counteract commercialism and the commodification of archaeological heritage, some archaeologists have attempted to preserve the scientific approach by tailoring interpretation to different audiences. In 1986, James constructed a model of archaeological interpretation with two levels: the first, internal and professional; the second, external or 'lay' (Robb, 1998). The growth in new dynamic, virtual and interactive technologies that allow for multiple audiences moves further along this path and has been accompanied by exciting developments in archaeological interpretation that acknowledge the importance of creativity and active visitor engagement in reconstructing the past. I will look at these issues now.

Immersive Media

'I am Vesuvius', thunders the voiceover in the Herculaneum Virtual Museum in Italy. Unlike traditional museum-based interpretations of Pompeii, this is the volcano's story of the folly of man in ignoring its power to destroy. The conceit is that nature has prevailed and will continue to prevail. The experience, viewed through 3D glasses, is frightening and convincing, so much so that I emerged looking fearfully at Vesuvius when I visited Pompeii.

Yet there is little of this excitement in the interpretation of Pompeii's ruins. Here we see a row of houses and understand who owned them and what happened within them, but not how the people felt or lived their lives; there we encounter information about how the town was configured spatially. The visitor infers the human drama of the eruption from the juxtaposition of the ruined buildings with what we know came later: the burying of both cities by ash, which hardened so much that it left impressions of people's bodies preserved in their final poses at the moments of death. This tragedy has been monetised by a series of blockbuster touring exhibitions, but the disjunction between the site and the mediated experience remains.

What would happen if social history were introduced on site? There is only one eye-witness account of Vesuvius's eruption: that of Pliny the Younger, who witnessed the eruption from the ancient port of Misenum. His account was based on his own experience and the first-hand reports of people who had been with his uncle, Pliny the Elder, who bravely turned back to Pompeii after the first eruption to rescue a friend of his family but was killed when Vesuvius erupted again (Investigating the Past: Pompeii and Herculaneum). Pliny the Younger's account is compelling:

> Ashes were already falling, not as yet very thickly. I looked round: a dense black cloud was coming up behind us, spreading over the earth like a flood. 'Let us leave the road while we can still see', I said, or we shall be knocked down and trampled underfoot in the dark by the crowd behind. We had scarcely sat down to rest when darkness fell, not the dark of a moonless or cloudy night, but as if the lamp had been put out in a closed room.

> You could hear the shrieks of women, the wailing of infants, and the shouting of men; some were calling their parents, others their children or their wives, trying to recognize them by their voices. People bewailed their own fate or that of their relatives, and there were some who prayed for death in their terror of dying. Many besought the aid of the gods, but still more imagined there were no gods left, and that the universe was plunged into eternal darkness for evermore. (Pliny in his second letter to Tacitus, Eyewitness to History, A.D. 79)

Although Pliny the Younger's eyewitness account forms the core of interpretation in the virtual museum at Herculaneum, a city which itself was swamped by hot mud and lava, an account of the terrifying ordeal to which Pliny the Elder succumbed on his second visit to Pompeii after the eruption was nowhere evident when I strolled around the ruins of Pompeii in 2017. Recently, however, Pompeii has offered a range of interpretive initiatives. These include guided tours by guides from Region Campania, an educational program for schools and 'Campania by night: archaeology under the stars', where:

> …the splendour and beauty of the illuminated temples and buildings, along with the ability to relive the moments of daily life in a unique place through voices, noises and sounds, will accompany the visitor through an emotional itinerary from the ancient Porta Marina into the midst of history. (Archaeological Park of Pompeii, 2019: n.p.)

Pompeii's interpretation illustrates that the gulf between archaeological practice and interpretation, first identified in the 1980s, is still wide in some places. But digital media may offer a way to link tangible and intangible heritage on sites. Digital technology can take many forms, including 360° flythroughs, virtual reconstructions, sound and light shows, virtual reality, augmented reality, immersive reality and mixed reality experiences, some of which are delivered via smartphones and tablets (Han *et al.*, 2018). Naturally, this has been met with concerns by those archaeologists who are still convinced that fabric has inherent value, that nothing substitutes for the 'real thing' – contact with the material fabric of sites – and that virtual interpretations can create an illusion of authenticity (Bateman cited in Llobera, 2012). In the latter case, authenticity is understood to reside in direct contact between the visitor and the fabric of the site as uncovered by the archaeologist and is conveyed through approved 'interpretation-lite' records of direct field experience such as videos, notebooks, photographs, sketches and other forms of representation about a site (Tilley cited in Llobera, 2012).

Heritage professionals are also hyper-aware that intangible values, and community-owned representations of the past, may threaten a site's material culture. Former ICOMOS President Gustavo Araoz (Araoz, 2011: 58) argues that 'the dispersal of values between material and intangible vessels increasingly comes at the expense of the historic fabric of the place'. In practice, this means that unscrupulous organisations may use interpretation to justify full or partial removal, or demolition, of fabric. Tourism also has its part to play in destroying sites. Comer and Willems (2011) argue that the archaeological record has become increasingly compromised as tourist numbers have grown and that this has increasingly affected the historical and scientific values of four of the most famous World Heritage archaeological sites: Petra, Machu Picchu, Pompeii and Angkor. In the worst cases, wiring for sound and light shows involved cutting channels in ancient stonework, and the Tether of the Sun monument at Machu Picchu was chipped during filming for a beer commercial in 2000 (Comer & Willems, 2011).

But digital media does not necessarily sit in Satan's talons; in fact, it may belong in the hands of the opposite deity if done well. Digital media shares the same underlying premise as all other forms of interpretation: it is a means of conveying a story, and the strength of story and quality of scholarship is always what determines how good the interpretation is. However, rather than being seen as simply another communication device, digital media has often stood for 'bad' interpretation in opposition to the 'good' interpretation that went before. This is to ignore all the implicit biases and subjectivity of archaeology, as discussed earlier.

A hot spot for arguments about the role of digital media interpretation is 3D visualisation, which ranges from a simple use of digital technologies for visually reconstructing archaeological sites, at one end of the spectrum, to 360° flythroughs, games, VR, AR, Mixed and Immersive Reality, at the other. Virtual Reality has existed in the archaeological sphere since the 1990s, with one of the first case studies being The Tomb of Queen Nefertari (Karlsson, 2013). However, virtual technology has proved a source of constant concern for heritage practitioners and archaeologists alike. Given that reconstruction is already a loaded term in heritage circles because it involves 'a level of interpretive certainty which is largely unobtainable' (Watterson, 2015: 120), 3D reconstruction can seem even worse. The 2009 *London Charter* is an attempt to deal with the issue of when and how to use computer-based visualisation, but there is still a wide variance in the application and quality of virtual technologies. Alice Watterson (2015) points out that the idea of 'virtual archaeology', first coined by Reilly (1991), has simply led in some cases to traditional concepts of archaeology, including the dualism between tangible and intangible cultural heritage, being reinforced by digital technology. At the Palace of Versailles, for example, room after room offers 3D reconstructions of the building and grounds, but there is no interpretation of the French Revolution or Marie Antoinette to link the social history of the place to its built fabric.

Yet digital media can also have many advantages for archaeological sites. When done well, it can span the whole spectrum of visitor engagement, allowing the interpretive story to be told before, during and after a visit. This means that complex historical and scientific information can be conveyed while leaving the site relatively untouched by interpretive infrastructure or 'books on signs'. 3D reconstructions also meet the theoretical needs of modern heritage interpretation. They can be used to facilitate discussions between the archaeologist, the site and the archaeological record, to deal with complex human agency, to stir the emotions of the visitor and to create empathy with the actors of the past (Watterson, 2015). While this approach inevitably raises issues of subjectivity and the incomplete nature of the archaeological record, theorists such as Wheatley (2000) believe that avoiding aesthetic and personal experience in visualisation is 'irrational and misguided' and that subjectivity in visual work and field methods should be acknowledged and engaged with as 'a core dimension of our interpretive process and representation' (Watterson, 2015: 122). Further, in line with postmodern approaches to history, making the archaeological, research and analysis processes explicit for audiences allows them to engage actively with the interpretation and for the 'inevitable uncertainties' of the interpretive process to remain visible and intact (Watterson, 2015: 122).

Video games, either onsite or offsite, also offer intriguing possibilities for creating an immersive experience of the past and empowering people to explore, navigate and interpret the archaeological record (Graham, 2017). *Never Alone*, a first-of-its-kind video game based on traditional Iñupiaq stories, was launched in Canada in 2014 (Cook Inlet Tribal Council, 2017: 21). As an 'atmospheric puzzle performer' where the avatar moves through the game to solve practical problems, it allows gamers to uncover the 'amazing culture' of the Iñupiaq. The game was so successful that it was nominated for numerous awards and created job opportunities for tribal members. One reviewer wrote:

> [*Never Alone*] teaches that the preservation of history is its own reward, and proves that video games have as much right to facilitate that process as any other art form. (Evans-Thirlwell, 2014, cited in Cook Inlet Tribal Council, 2017: 28)

Copplestone (2017) found that creating games while archaeological excavations were taking place meant that players could play at being an archaeologist at the same time as reflecting critically on how archaeological knowledge had been constructed. Gaming makes the past tangible and visceral in a way that traditional media does not. By navigating around a virtual environment with body movements, for instance, users can directly interact with the inhabitants of archaeological sites (Karlsson, 2013). The current UNESCO Chair in Cultural Visualisation and Heritage, Erik Champion, theorises that this will allow the general public to 'imagine the situated cultural significance of that site as it was once viewed, understood and inhabited, offering an immersive way to link tangible and intangible heritage' (Champion, 2017: 116). While archaeogaming is still relatively new, future games will feature increasingly sophisticated interactive immersive visual worlds that acknowledge that data changes and technology changes and provide for different learning styles (Champion, 2017).

Citizen Archaeologists

Public archaeological digs have always been popular, but collaborative partnerships in gaming offer an exciting future for archaeological interpretation combined with a chance to fascinate young audiences with the idea of archaeology. In 2015, the 'Crafting the Past' project in Scotland engaged games-based learning specialists Immersive Minds to bury a Roman amphitheatre in a *Minecraft* world (McGraw *et al.*, 2017). Players from the online *Minecraft* gaming community were asked to excavate the site 'as an archaeologist would' in a virtual world that had been painstakingly recreated through research, site visits and mapping (McGraw *et al.*, 2017). Online participants immediately began undertaking research of their own and interrogating the site (as an archaeologist or historian would), asking questions such as: 'Where are the toilets on a dig site?' and 'Did the Romans actually reach Scotland?' (McGraw *et al.*, 2017: 169). The team then created a second virtual dig at Watling Lodge, a Roman site along the Antonine Wall in Scotland, which took place at the same time as a live dig. At this virtual dig, 'children used digital tools to uncover the *Minecraft* build and real archaeological tools to uncover the past in the real world' (McGraw *et al.*, 2017: 173). A member of the Immersive Minds team also acted as a Roman ghost and answered children's questions.

Historical *Minecraft* games have proven a massive hit with young people in Scotland, particularly the disengaged 16–24 age group, with some Young Archaeologist Clubs now using *Minecraft* to build their own heritage recreations and archaeological digs (McGraw *et al.*, 2017). Gaming has also opened up the field of archaeology for both gaming participants and game creators, making them new players in archaeological interpretation: a field that 'often struggles to tell (stories) effectively' in the words of the 'Crafting the Past' team (McGraw *et al.*, 2017: 179).

Conclusion

This chapter has outlined broad changes in the fields of archaeology and heritage interpretation that have affected the practice of archaeological interpretation. Some recent developments indicate that the fields of archaeology and heritage interpretation are drawing more closely together professionally. In 2007, for instance, the ICOMOS International Scientific Committee for the Interpretation and Presentation of Cultural Heritage Sites (ICIP) was established by a Committee that included several noted archaeologists. Nine years later, the ICOMOS 2016 symposium on post-disaster reconstruction was extremely popular and raised many important theoretical issues about how digital technologies affect the conservation, preservation and interpretation of archaeological sites.

In 2008, Francis McManamon (2008: 458) stated that 'too many professional archaeologists are still uninterested in public education and outreach'. Times have changed. Methodologies from participatory interpretation, community-based research and pro-poor tourism are not only reshaping archaeological practice but also providing exciting new outcomes for public archaeological work. At the same time, the extraordinarily rapid developments in digital technology over the last 10 years have changed the interpretation of archaeology forever. While all of these developments remain contested and debated within the archaeological and heritage professions, they have brought archaeology to the forefront of public life in ways never experienced before. The future is bright. To end with the words of Spinal Tap, archaeological interpretation is poised to 'Go to 11'.

References

Ablett, P.G. and Dyer, P.K. (2009) Heritage and hermeneutics: Towards a broader interpretation of interpretation. *Current Issues in Tourism* 12 (3), 209–233.

Araoz, G. (2011) Preserving heritage places under a new paradigm. *Journal of Cultural Heritage Management and Sustainable Development* 1 (1), 55–60.

Archaeological Park of Pompeii (2019) Pompeii: Visit the Archaeological Sites. See www.pompeiisites.org (accessed 28 January 2019).

Barry, K.M. (2014) Framing the ancients: A global study in archaeological and historic site interpretation. Unpublished doctoral dissertation, College of Arts and Architecture, the Pennsylvania State University.

Binks, G. (1986) The interpretation of historic monuments: Some current issues. In M. Hughes and L. Rowley (eds) *The Management and Presentation of Field Monuments* (pp. 39–46). Oxford: Oxford University Press.

Champion, E.M. (2017) Single white looter: Have whip, will travel. In A.A.A. Mol, C.E. Ariese-Vandemeulebroucke, K.H.J. Boom and A. Politopoulos (eds) *The Interactive Past: Archaeology, Heritage and Video Games* (pp. 107–122). Leiden: Sidestone Press.

Cobb, H. and Croucher, K. (2014) Assembling archaeological pedagogy: A theoretical framework for valuing pedagogy in archaeological interpretation and practice. *Archaeological Dialogues* 21 (2), 197–216.

Comer, D.C. and Willems, W.J.H. (2011) Tourism and archaeological heritage: Driver to development or destruction? Presented at ICOMOS Paris 2011.

Cook Inlet Tribal Council (2017) Storytelling for the next generation: How a non-profit in Alaska harnessed the power of video games to share and celebrate cultures. In A.A.A. Mol, C.E. Ariese-Vandemeulebroucke, K.H.J. Boom, H.J. Krijn and A. Politopoulos (eds) *The Interactive Past: Archaeology, Heritage and Video Games* (pp. 21–32). Leiden: Sidestone Press

Copplestone, T.J. (2017) Designing and developing a playful past in video games. In A.A.A. Mol, C.E. Ariese-Vandemeulebroucke, K.H.J. Boom, H.J. Krijn and A. Politopoulos (eds) *The Interactive Past: Archaeology, Heritage and Video Games* (pp. 85–97). Leiden: Sidestone Press.

Downs, J., Latham, P., Sheppard, E. and Sheppard, P.A. (2009) *Investigating the Past: Pompeii and Herculaneum*. (Video Production) Mona Vale, NSW: Phil Sheppard Video Productions.

Evans-Thirlwell, E. (2014) Never Alone review. *PC Gamer*. See https://www.pcgamer.com/never-alone-review/ (accessed 20 January 2019).

Fowler, D.D. (1987) Uses of the past: Archaeology in the service of the state. *American Antiquity* 52 (2), 229–248.

Giblin, J. (2015) Archaeological ethics and violence in post-genocide Rwanda. In A. Gonzáles and G. Moshenska (eds) *Ethic and the Archaeology of Violence* (pp. 33–49). New York: Springer.

Gonzalez-Tennant, E. (2011) Archaeological Research and Public Knowledge: New Media Methods for Public Archaeology in Rosewood, Florida. Unpublished doctoral dissertation, University of Florida.

Graham, S. (2017) On games that play themselves: Agent based models, archaeogaming, and the useful deaths of digital Romans. In A.A.A. Mol, C.E. Ariese-Vandemeulebroucke, K.H.J. Boom and A. Politopoulos (eds) *The Interactive Past: Archaeology, Heritage and Video Games* (pp. 123–131). Leiden: Sidestone Press.

Guest, C., Shearer, H., McKean, M. and Reiner, R. (1984) *Stonehenge* lyrics ©. New York: Sony Music Publishing.

Han, D.-I., tom Dieck, M.C. and Jung, T. (2018) User experience model for augmented reality applications in urban heritage tourism. *Journal of Heritage Tourism* 13 (1), 46–61.

Hodder, I. (1991) Interpretive archaeology and its role. *American Antiquity* 56 (1), 7–18.

Hoffman, T.L., Kwas, M.L. and Silverman, H. (2002) Heritage tourism and public archaeology. *The SAA Archaeological Record* 2 (2), 30–32.

James, N. (1986) Leaving it to the experts. In M. Hughes and L. Rowley (eds) *The Management and Presentation of Field Monuments* (pp. 47–58). Oxford: Oxford University Press.

Jameson, J.H. (1997) *Presenting Archaeology to the Public: Digging for Truths*. Walnut Creek, CA: AltaMira.

Jameson, J.H. (2007) Interpretation of archaeology for the public. In D.M. Pearsall (ed.) *Encyclopaedia of Archaeology* (pp. 1529–1543). Amsterdam: Elsevier.

Karlsson, C. (2013) *Visualising Archaeology with Virtual Reality Tools*. Unpublished master's thesis, Department of Archaeology and Ancient History, Lund University, Sweden.

Karskens, G. (1999) *Inside the Rocks: The Archaeology of a Neighbourhood*. Sydney: Hale & Iremonger.

Llobera, M. (2012) Life on a pixel: Challenges in the development of digital methods within an 'interpretive' landscape archaeology framework. *Journal of Archaeological Method Theory* 19 (4), 495–509.

McGimsey, C.R. (1972) *Public Archaeology*. London: Seminar Press.

McGraw, J., Reid, S. and Sanders, J. (2017) Crafting the past: Unlocking new audiences. In A.A.A. Mol, C.E. Ariese-Vandemeulebroucke, K.H.J. Boom and A. Politopoulos (eds) *The Interactive Past: Archaeology, Heritage & Video Games* (pp. 167–184). Leiden: Sidestone Press.

McManamon, F.P. (2008) Archaeological messages and messengers. In G. Fairclough, R. Harrison, J. Jameson and J. Schofield (eds) *The Heritage Reader* (pp. 457–481). London: Routledge.

Muir, J. (1912) *The Yosemite*. Garden City, NY: Doubleday.

Nagtegaal, R. (2015) Walk my Street to Rome: Community Perceptions of Roman Heritage. Unpublished MLE thesis, Department of Environmental Sciences, Wageningen University, Netherlands.

Opp, J. (2011) Public archaeology and the fragments of place: Archaeology, history and heritage site development in southern Alberta. *Rethinking History* 15 (2), 241–267.

Pishief, E. (2017) Engaging with Māori and archaeologists: Heritage theory and practice in Āotearora New Zealand. In B. Onciul, M.L. Stefano and S. Hawke (eds) *Engaging Heritage, Engaging Communities* (pp. 55–72). Woodbridge: Boydell Press.

Reilly, P. (1991) Towards a virtual archaeology. In K. Lockyear and S. Rahtz (eds) *CAA90: Computer Applications and Quantitative Methods in Archaeology* (pp. 133–139). Oxford: British Archaeological Reports.

Robb, J.G. (1998) Tourism and legends: Archaeology of heritage. *Annals of Tourism Research* 25 (3), 579–596.

Saitta, D.J. (1994) Agency, class and archaeological interpretation. *Journal of Anthropological Archaeology* 13 (3), 201–227.

Schorch, P. (2017) Assembling communities: Curatorial practices, material cultures and meanings. In B. Onciul, M.L. Stefano and S. Hawke (eds) *Engaging Heritage, Engaging Communities* (pp. 31–46). Woodbridge: Boydell Press.

Silberman, N.A. (2013) *Heritage Interpretation as Public Discourse: Towards a New Paradigm. Selected Works of Neil A. Silberman*. Amhurst: University of Massachusetts.

Smith, L. (2008) Politics of archaeology. In D.M. Pearsall (ed.) *Encyclopaedia of Archaeology* (pp. 1853–1855). Amsterdam: Elsevier.

SPAB (2019) The SPAB Manifesto. See https://www.spab.org.uk/about-us/spab-manifesto (accessed 30 January 2019).

Stone, P. and MacKenzie, R. (1989) Is there an excluded past in education? In D.L. Uzzell (ed.) *Heritage Interpretation*, Vol. 1 (pp. 113–120). London: Belhaven.

Tilden, F. (1957) *Interpreting Our Heritage*. Chapel Hill, NC: University of North Carolina Press.

Tunbridge, J. (2017) Interview – John Tunbridge. In B. Onciul, M.L. Stefano and S. Hawke (eds) *Engaging Heritage, Engaging Communities* (pp. 47–50). Woodbridge: Boydell Press.

Watterson, A. (2015) Beyond digital dwelling: Re-thinking interpretive visualisations in archaeology. *Open Archaeology* 1, 119–130.

Wheatley, D. (2000) Spatial technology and archaeological theory revisited. In K. Lockyear, T. Sly and V. Mihailescu-Birlitsa (eds) *Computer Applications and Quantitative Methods in Archaeology* (pp. 123–131). Oxford: BAR International Series.

Wickham-Jones, C.R. (1988) The road to Heritat: Archaeologists and interpretation. *Archaeology Review Cambridge* 7, 185–193.

Widner Ward, C. and Wilkinson, A.E. (2006) *Conducting Meaningful Interpretation: A Field Guide for Success*. Golden, CO: Fulcrum Publishing.

Young, C. (2013) An international view of reconstruction. In N. Mills (ed.) *Presenting the Romans: Interpreting the Frontiers of the Roman Empire World Heritage Site* (pp. 75–83). Woodbridge: Boydell Press.

12 Archaeology, Nationalism and Politics: The Need for Tourism

Gai Jorayev

Introduction

This chapter aims to tackle three themes: archaeology, nationalism/politics and tourism. These themes regularly overlap given the realities of archaeological heritage management in the 21st century, and therefore their interaction demands a broader examination. Within the context of this volume, this chapter provides an overview of the state of the scholarship on that interaction and aims to be a literature-rich synopsis of the junction of the three themes, albeit with certain references to wider examples to highlight the points made. Constraints of length demand only a selective gaze to the literature and sources, which are by now very extensive in these three areas. The chapter also looks at these in an international context, with involvement of broader themes such as World Heritage and global tourism.

Archaeology and Nationalism

The use of archaeology for the purposes of nation-building or nationalism is not a new phenomenon, as Chapter 11 illustrates. Cases from different parts of the world are plentiful in modern literature (e.g. Champion & Díaz-Andreu, 1996; Fowler, 1987; Knapp & Antoniadou, 1998; Kohl, 1998; Kohl & Fawcett, 1996; Kohl *et al.*, 2007; Meskell, 1998, 2002; Newell, 2008; Shnirelman, 2009; Trigger, 1984), and archaeology, especially archaeological interpretations, are expected to play an important role in strengthening national identity and revealing the narratives of the people in power. The study of nationalism in the last few decades is rich and influenced by several major theories (Hutchinson & Smith, 1994).

Four key works (i.e. Anderson, 1983; Gellner, 1983; Hobsbawm & Ranger, 1983; Smith, 1991) and their later editions influenced research on nationalism as part of many disciplines, including archaeology and heritage studies (Askew, 2010; Winter, 2012). It is commonly accepted that the

1980s and 1990s saw increased public and academic attention to nationalism studies because of the rise of nationalism in the European socialist bloc and the subsequent collapse of state socialism, which served as a catalyst for researching nationalism in discussions of nation-building (Connor, 1993; Hutchinson & Smith, 1996; Kohl et al., 2007; Suny, 2001; Treanor, 1997). Since then the leading works of nationalism studies have been subjected to a fair amount of critique (Connor, 1993; Guibernau, 2004; Kaufmann, 2017; O'Leary, 1997; Wogan, 2001), but their influence remains strong. The distinctions made by Smith (1991) between ethnic and civic nationalism in particular still play a significant role in explaining the interactions between nationalism and archaeology.

As is often highlighted, archaeology's relationship with, and dependency on, the state, particularly in funding, creates a foundation for its involvement in narratives of nationalism, which are in turn driven by states' desire to create long and glorious origin myths (Sommer, 2017). If we take the view that 'heritage is a primary instrument in the "discovery" or creation and subsequent nurturing of national identity' (Graham et al., 2005: 29) and that 'nationalism' grows together with the concept of 'the nation', then it is almost inevitable that nation-states will seek to find a role for the archaeological record in their national(istic) narratives. Several important studies put the history of archaeology and the development of nationalism together and look at the reasons, justifications and outcomes (Atkinson et al., 1996; Díaz-Andreu García, 2007; Díaz-Andreu García & Champion, 2015; Kohl, 1998; Kohl & Fawcett, 1996), and as Tim Winter (2015: 331) notes, 'the coupling of a material culture of the deep past with the politics of nationalism and the making of national citizens remains as vibrant, and in some cases as troubling, as ever'.

However, assumptions should not be made that archaeology is almost destined to become a manipulation tool under a strong nationalistic state system; important research findings show that scientific archaeological research takes longer than short waves of intense nationalism and therefore can stay objective (Galaty & Watkinson, 2004; Junker, 1998; Sommer, 2017). It is also very clear that it is not always straightforward to use Western concepts of ethnicity or nationalism in analysing different regions of the complex modern world (Eriksen, 1993: 16). Although the narratives of nationhood are indeed difficult to displace once they are established (Sommer, 2017: 183), they are not very easy to establish in the first place in the age of information and rapid global shifts.

The Role of Archaeology in Society

Archaeology and nationalism are often argued to be useful for the ideologies of nation-states, though there are other immediate worries for a state in a modern setting that can at times take a higher priority to ideology or origin myth. Especially in countries undergoing socioeconomic changes, governments are often elected on promises of economic prosperity or wellbeing, but rarely on their promise of providing better support for the nation's past. However, archaeology and the broader heritage narrative are expected to play their role in economic and social development, as if in a more blunt modern representation of Gellner's (2008) theoretical consideration of the interactions of culture, community and modern economy of a state. Positive valorisation of the remains of the past is seen as one of the factors behind the professionalisation of heritage institutions in the first place (Carman & Sørensen, 2009), and tangible heritage is recognised as an economic asset alongside its cultural values in international normative texts and guidelines for its potential contribution to development (Cernea, 2001; Council of Europe, 2005; UNESCO WHC, 2011). This notion is closely aligned with the overarching culture and (sustainable) development narrative globally (Bandarin *et al.*, 2011; UNESCO, 2017a, 2017b) and many expectations are placed on cultural heritage in the age of globalisation (Labadi & Long, 2010).

It seems that from earlier discussions of the overall justifications for archaeology (Shanks & Tilley, 1992), the conversation has now firmly moved to more focused debates that look at archaeological heritage and its meanings for society, although listing them all would be an impossible task. There are some valuable recent discussions specifically in the area where the economy and archaeology come face-to-face. For instance, Flatman (2012) discusses the relevance of archaeology to society under the harsh realities of modern times. Burtenshaw (2014, 2017) looks at bridging the gap between the economic and cultural values of archaeology and the role of economics in public archaeology. Klamer (2014) considers including financial value into the evaluation of cultural heritage sites, and Gestrich (2011) asks if it is ever a good idea to think about the monetary value of archaeology. But detailed reviews of the economic impact of heritage on specific sites are still few (see Chapters 2 and 3), despite this being on the research agenda for some time (Timothy & Boyd, 2006; VanBlarcom & Kayahan, 2011). Similarly, detailed studies on the benefits of World Heritage status to local economies (e.g. Rebanks Consulting, 2009; VanBlarcom & Kayahan, 2011) remain few; cases demonstrating both the economic benefits of World Heritage status and its lack of benefits are also rare (Lyon & Wells, 2012). It is also true that, although growing, studies on the impact of major economic shifts on the profession of archaeology remain few (Aitchison, 2009, 2015; Lennox, 2018; Rock-Macqueen, 2018).

However, what is noticeable is that organisations and professionals working in the fields of culture and cultural heritage are getting better at quantifying or assessing the benefits of their work in a manner that is understandable for a wider group of political decision makers. It is interesting to observe the change of language used to justify the overall role of culture and the arts in the UK (Arts Council England, 2012, 2013, 2014; CEBR, 2013, 2015; DCMS, 2015; O'Brien, 2010) where the discussion is becoming more based on the economic advantages of heritage assets, including archaeological remains, under recent government austerity policies, which among other things oversaw the split of English Heritage into two organisations for economic reasons (Larkin, 2014).

Archaeological Heritage as a Driver of Tourism and Tourism as an Economic Driver

The expectation of archaeology contributing to local economies in the form of tourism revenue exists almost within a state of default inertia. A vast array of assets is brought together under the umbrella of cultural heritage and expected to be mobilised to enable cultural tourism, heritage tourism, archaeotourism or simply tourism. Despite efforts by archaeologists and other heritage specialists to demonstrate and explain complex benefits to be gained from the use of heritage – in terms of education, local regeneration, local identities and societal cohesion, in the author's experience discussing long-term management planning with local authorities in Asia, Africa and Europe, tourism has a tendency to arise as one of the first considerations.

This is not surprising, as decision makers often require 'ideas of value' to inform their decisions (Burtenshaw, 2017: 33), and tourism is an area where the numbers, be it in the form of incoming visitors or amounts spent per night, can be generated to measure that value. There are significant flaws in these assumptions as the numbers and the statistics generated by the heritage tourism sector may act as indicators of failure, as well as indicators of success. The numbers alone can be seen as indicators of destruction or successful adaptive reuse. High visitor numbers can damage the heritage asset in question (see Chapters 9 and 10) or increase the requirement for additional expenditures for infrastructure and conservation, and therefore might negate the economic gains from visitors. Additionally, decoupling cultural or archaeological tourists from other categories is not easy owing to the complexities associated with such measurements (e.g. Cuccia & Rizzo, 2013). The experiences of the author in the developing world suggest that it is often the case that authorities with opaque statistics and economic calculations in the first place are more likely to be willing to foster archaeological tourism as they underestimate the potential for failure.

The expectations are often fuelled by growth figures of international tourism, which has seen a strong upward trajectory in the recent past, despite major safety concerns, protests against 'overtourism' and natural disasters (ITB Berlin, 2015, 2016, 2018). In fact, international tourist arrivals have been growing annually since 2010 (UNWTO, 2017b) and reached their highest levels for seven years (+7%) in 2017 (UNWTO, 2018), to the extent that even the World Tourism Organization needed to reassure that if managed in sustainable ways, the growth of an industry that is estimated to provide 10% of the world's GDP is not a problem (Rifai, 2017). Within that upward trajectory, the demand for archaeological heritage-based tourism is increasing too (UNWTO, 2016b).

The relationships between tangible and intangible heritage and the behemoth tourism industry are complex and multi-faceted (Park, 2013; Salazar & Zhu, 2015; Timothy & Boyd, 2006), and the multitude of its dimensions are at the core of international agreements (ICOMOS, 1999). Equally, as it is the subject of major academic works, the relationship between cultural/heritage tourism and society is complex (Herbert, 1995; Lyon & Wells, 2012; Smith, 2003; Timothy & Boyd, 2003). How tourism is seen from a heritage studies point of view is changing rapidly (Winter, 2010). The major international bodies – UNESCO, UNWTO, ICOMOS among them – consider it their duty to produce handbooks and manuals related to managing tourism and its complexity at archaeological sites in general and at World Heritage Sites (WHSs) in particular (ICOMOS & UNWTO, 1996; Pedersen, 2002; UNWTO, 2004).

World Heritage, Nationalisms and Tourism

Assumptions are often made that the recognition of the outstanding value of archaeological heritage, particularly in the form of World Heritage Sites, almost automatically leads to an increase in tourism (Frey & Steiner, 2011; Timothy, 2014) or, conversely, that governments seek to promote World Heritage because of the potential tourism benefits. With the current diversity of the UNESCO Word Heritage List and the obvious problematics with quantifying the impact of such a recognition, this issue is complicated and requires greater consideration beyond the simple assumptions outlined above. Some accounts indeed show that World Heritage listing does not necessarily increase tourist flows on its own (Cellini, 2011; Hall & Piggin, 2003), sometimes leaves very little economic impact locally if not managed well (Orbaşli, 2013), and that the outcomes can be mixed (Jimura, 2011; Lee *et al.*, 2018).

But it is exactly the World Heritage listing that is seen as symptomatic of tourism's development agendas, as well as the influence of nationalism (Labadi & Long, 2010). If one takes the view that heritage is being turned into a commodity, then '"World Heritage" was the most marketable of this form of commodity' (Harrison, 2013: 89). Additionally, there is a strong argument that despite universalist agendas and the involvement of good-intentioned practitioners, UNESCO's World Heritage system and 'the globalised and institutionalised heritage system has not overcome nation-state based power structures and nationalist agendas, but has rather enhanced them' (Askew, 2010: 20). Although the agendas of tourism development also lead state party efforts to nominate more cultural assets as World Heritage, the experience of recent times suggests that the agendas of nationalism – or nation-building, nation-branding and regional competition – dictate the efforts of the states even more (Meskell, 2015).

There are current and recent examples of World Heritage inscriptions seething of nationalism. One that stands out is the Temple of Preah Vihear WHS, the nomination of which raised considerable tensions between Thailand and Cambodia (Williams, 2011; Winter, 2010). A site with clear tourism potential became a symbol of insecurity very quickly when given World Heritage status, and it is hard to suggest that UNESCO was unaware of or irresponsive to the Preah Vihear developments (World Heritage Centre, 2011). A deeper and more nuanced look into the temple (Pawakapan, 2013) reveals that the fervours of nationalism agitated around the concept of heritage can be very contradictory to the state's previous positions, detrimental to its overall goals and often expressed by specific groups rather than the totality of the nation-state.

There are other examples where archaeological heritage in general, and World Heritage in particular, stir up nationalistic discourses (Trigger, 1984; Winter, 2012). The entire recent episode of a diplomatic spat between the USA, Israel and Palestine being played out in the UNESCO forum, which resulted in a full withdrawal of the United States from UNESCO membership on 1 January 2019, is an even larger example of interactions of archaeological heritage, nationalism(s) and politically charged international relations (Beaumont, 2017; Coningham, 2017; Lynch, 2017). An argument could be made that this sad saga almost deliberately tries to ignore that UNESCO is a member state-based UN body and not an independent decision-making organisation. The World Heritage listing process in fact turns this branch of the United Nations that is tasked to ensure peace by protecting and sharing culture among other things, into a centre point of these culture/heritage-based nationalisms. Perhaps this is the ultimate proof that culture and heritage, including archaeological heritage, is always likely to remain influenced, if not consumed, by nationalistic overtones throughout the globe (Champion & Díaz-Andreu, 1996; Kohl & Fawcett, 1996; Meskell, 1998; Sørensen, 1996).

The discussions over the role of UNESCO and its World Heritage List, its relevance or irrelevance, its practicality or absurdity, have been ongoing in academic discourse for some time (Anglin, 2008; Askew, 2010; Frey & Steiner, 2011; Harrison, 2013; Maurel, 2017; Norman, 2011), and it is not the intention here to delve specifically into the broad field of the values of the World Heritage system. Scepticism should be expressed though, that tourism and its economic potential underpins the majority of World Heritage nominations. These are complex processes, and it is hard to gauge the true motivations of state parties (Meskell, 2015); seemingly, tourism is often an afterthought despite WHSs frequently being seen as 'branding' tools for promoting tourism to certain destinations (Labadi & Long, 2010). There are of course numerous heritage sites that would be tourist magnets regardless of their recognition by international bodies. Some sites may have an element of 'dutiful tourism', as defined by Hughes

(2008), even without the UNESCO brand. Likewise, some WHSs might not be 'must-see' attractions for everyone (Adler, 1989; Edensor, 2001).

Evolution of the Sustainability Concept

Sustainable tourism is a key phrase in international discussions of cultural heritage (Robinson & Picard, 2006; UNCTAD, 2013; United Nations, 2016; UNWTO, 1997), as well as among individual scholars and practitioners (Butler, 1999; Girard & Nijkamp, 2009; Hall & Richards, 2000; Mowforth & Munt, 2016). Many are indeed well aware of the complex nature of heritage tourism and its requirement to be a careful balancing act.

Discussions of sustainability are taking place in much wider areas than before and include issues of poverty alleviation and support for the most disadvantaged communities (Anderson, 2015; Roe & Urquhart, 2002). Sustainability is also important at the global level of tourism operations, partly because of a rapid change in travellers' expectations and partly because of its potential in market positioning (Bender, 2013; Butcher, 2003; Nickerson *et al.*, 2016; Pulido-Fernández & López-Sánchez, 2016; Weaver, 2012). However, identifying a clear route to sustainability in heritage tourism, with it taking place in so many settings and environments, is as hard as finding sustainability in heritage management itself.

As mentioned above, cultural heritage and development is a popular topic within the global 'sustainable development' narrative (Mergos & Patsavos, 2017; UN, 2018). This is part of the wider 'culture for development' agenda (Bandarin *et al.*, 2011; Schech & Haggis, 2000; UNESCO, 2017a, 2017b) and is firmly becoming part of the 'culture and creative industries' concept, at least in Europe (European Commission, 2018; KMU Forschung Austria & VVA Europe, 2016). Heritage and tourism also often involve the term 'development' before turning to economics (Timothy, 2014); however, it is important to frame development not only as economic in discussions of heritage tourism. The economic impact of tourism for the localities involved is questionable in many cases, and the desirability of tourism as a strategy for economic development is frequently debated (Urry & Larsen, 2011), while the economic commodification of heritage in general is often, and extensively, critiqued in heritage studies literature (Smith, 2006; Timothy, 2011). There are clear benefits of tourism beyond its economic power, and some key international texts rightly highlight its potential for cultural exchange (ICOMOS, 1999). The economic advantage is not always the easiest to gain in locally beneficial ways. Although heritage tourism might serve as an opportunity for attracting investments to specific sites, even that could be a double-edged sword with tourism leakage being such a significant issue in many parts of the world (Lange, 2011; UNCTAD, 2013; Wood, 2017). The challenges of heritage tourism in broader terms are not new, although newer and more rapid forms of those challenges are appearing. Issues associated with the rapid increase of mass tourism are well documented and often find reflections in mass media as a running commentary (e.g. Connolly, 2017; Kettle, 2017; Peter, 2017) and in academic publications with a more balanced perspective (Butcher, 2003; Fletcher *et al.*, 2017; Urry & Larsen, 2011).

Sustainability is a key word in archaeological management too, but as Burtenshaw and Palmer (2014: 23) point out so well, sustainable heritage initiatives not only require understanding of archaeology in a given location, but also of a complex set of current conditions. Archaeologists may often be well-aware of those additional issues and conditions, but they may not be the ones in charge of developing the often economics-driven tourism packages. Closer inter-agency collaboration and involvement is an increasingly obvious path and should be seen as part of the holistic management of archaeological sites.

Archaeology and Tourism: Is it so Different from the Rest?

Although tourism to archaeological sites manifests some differences from other cultural tourism types (Willems & Dunning, 2015), maybe the divide between archaeotourism and cultural tourism is an unnecessary one. But on the other hand, there are tourism planning issues, such as accessibility and concerns about carrying capacities that may be unique to archaeological areas. This is the case not only when the sites are made 'display quality', but also during the process of archaeology itself as a tourist attraction.

The fragility of archaeological remains is an overarching concern in the process, but just thinking about the damage to the archaeological fabric is not sufficient, as carrying capacity is also 'about the tolerance levels of the tourists' (Fletcher *et al.*, 2017: 233). There are several texts that act as best practice guidelines for tour companies, site managers and even for the tourists themselves (AIA, 2017; UNWTO, 1997). Using archaeological resources for tourism almost inevitably erodes or changes them, but using them may make life-changing improvements for the locality and its population if managed properly. It can, in fact, play a political role in empowering communities politically, socially and economically.

The importance of interpretation is also paramount in enabling tourism at archaeological sites and in diffusing the tensions raised by fervours of nationalism. Heritage interpretation in general is an area of rich literary foundations both old (Tilden, 1957) and new (Lehnes & Carter, 2016), including the ones with support from official bodies (Hems & Blockley, 2006). There is also a large and valuable discussion on museum interpretations of heritage (Corsane, 2005; Kirshenblatt-Gimblett, 1998; Macdonald, 2010; Moser, 2010; Moussouri, 2014; Pearce, 1994; Roberts, 1997), which is very relevant. Archaeology and education as a separate area of scholarship also can make huge contributions to this discussion (Corbishley, 2011). Explaining heritage protection as part of visitor interpretation (Willems & Dunning, 2015) and turning the elements of contemporary life into constituent parts of the interpretation (ICOMOS, 1999) provide not only a modern-day context to the archaeological resources, but also raises awareness of the needs of local communities. Heritage sites are not static, and changes are not always positive, but change itself can be presented to explain recent history and add significance. That proposition depends on the skill and understanding of the specialists at both ends of the tourism market: companies selling the tours and staff on the ground, such as guides. Heritage specialists and archaeologists can make important interpretative contributions to that. Interpretation makes the sites significant, and it is interpretive stories, rather than photographic opportunities, that evoke feelings.

Heritage is a resource, but crucially, it is not simply an economic resource (Graham et al., 2005); better interpretation and participation may help explain that. Modern approaches to public archaeology and community engagement encourage archaeologists to impact local development agendas (Burtenshaw & Palmer, 2014) and tourism and community participation are highly regarded terms in the management of WHSs (Rasoolimanesh & Jaafar, 2017; Su & Wall, 2012). But the participation of the public and local communities in archaeology is multi-faceted and complex (Thomas & Lea, 2014) and often comes with its own sensitivities and tensions. Sustainability and community engagement are both relevant themes from a tourism angle (Hall & Richards, 2000), as well as approaches to archaeology and heritage management, with a significant overlap provided by publications looking at those in combination (Girard & Nijkamp, 2009). Direct benefits for the host communities are widely seen as part of the sustainable management of heritage, and as recent in-depth studies show (Dragouni, 2017), community involvement for sustainable heritage tourism can be effective.

Conclusions and Problematics

Heritage, politics, nationalism and tourism can be interlinked in a complex web of relationships in many contexts (Kaminski et al., 2013; Park, 2013; Timothy, 2011, 2014), and with the focus on archaeological heritage-based tourism, this contribution provided another look at the crossover between tourism and archaeology with the addition of a political and nationalistic perspective. The concept of heritage is interactive and dynamic (Salazar & Zhu, 2015) and can therefore be central in developing both tourism and nationalism.

The interaction of archaeology and nationalism are well researched, and similarly, the connection of archaeology and tourism is also well discussed in the literature. The feeling one may get from the literature is that archaeology and nationalism research, and archaeology and tourism research are sometimes done by different groups and with different frameworks being employed. It is not surprising given that scholars from many fields – anthropology, cultural heritage, museum studies, sociology, international relations, nationalism, development, economics and tourism studies to name the main ones – look at these interactions. Tourism studies itself is also a large field that draws upon many fields such as economics, sociology and geography (Adler, 1989). The languages used are not always identical; the extent of discussion is very different, and elaboration (or over-elaboration) of the same issues is not always easy to correlate. They may seem to look to the same concepts from very different academically grounded points of view. Also, while archaeology and nationalism seem to be researched mostly post-factum from a historical perspective, the fields of archaeology and tourism are constantly evolving and fast moving. Even this rather limited look at the literature suggests that our understanding of the interaction between archaeology, nationalism and tourism is quite rich, especially when it comes to archaeology and tourism. However, the recurring nature of the same problems over the last 3–4 decades in the case study-based literature also suggests that our ability to make game-changing contributions to the practical world of heritage management is rather limited. This may not be because of a lack of effort by heritage practitioners and can be related to overarching agendas of nation-states, but it is nevertheless worrying.

The role of the state is paramount in all of these issues. Be it localised management and research of archaeology, international interactions via the UN bodies such as UNESCO over heritage, or the development of heritage tourism, the state is often the responsible party and ultimate 'owner', 'representative' or 'investor'. States are often the creators of legislative and normative tools to ensure research, protection, management and valorisation of heritage. But very often, the states are also ultimate framers of nationalisms and therefore the search for heritage assets neutral of nationalist values or political agendas is perhaps fruitless from a philosophical point of view. Archaeology as a storytelling (as well as myth-making) practice has enormous value both for nationalisms (in their different forms or grades), as well as for tourism where the stories make the products (the sites or objects) attractive. The archaeological research process itself, not only its findings, also has become a key area of interest for nationalist interpretations. Involvement of international institutions to carry out archaeological investigations also could add to the argument of a recognition of the nation-state and its heritage policies.

The role of tourism as an employment provider and generator of much-needed income is well established, and current statistics also confirm that the industry employs one in 10 on the planet (UNWTO, 2017a). The economic benefits of tourism are clear, as well as archaeology's ability to attract visitors. However, putting well-developed international tourism into place is often beyond the state alone, especially if the state is isolationist, which it is likely to be if it follows strict nationalistic agendas. It is true that even the most isolationist regimes may want to facilitate tourism for specific purposes (Kim *et al.*, 2007; Wang *et al.*, 2017), and archaeological heritage can feature strongly in that. However, nationalisms, in their stronger incarnations, are often against the influx of 'others', with tourists being part of the other. Similarly, people may be less inclined to visit countries that are low on perceptions of inclusivity and friendliness.

On a very practical level, it is also often puzzling that despite some well-documented cases against it, governments will seek higher numbers of visitors rather than more sustainably and ethically minded lower numbers. There is an opportunity to discuss the relationship between nationalism and transparency in general. Although difficult to test empirically, it is likely to be the case that the states with strong nationalistic governments are also likely to have opaque or unclear data, which might result in unrealistic expectations of the potential for heritage tourism. Expecting the nationalist regimes to protect and promote the archaeological heritage will be short-sighted of course. However, proper nationalism often requires a strong state system and bureaucracy and the resulting inertia may also provide a strong state heritage management system. It is very popular in academic literature to highlight the areas and periods when heritage assets were selectively interpreted, but that does not necessarily mean they were also selectively protected. Protection is often a blanket approach, while the promotion of certain sites is often political.

As in many other fields, technology is appearing as a disruptive force. It has the potential to enable wider participation in discussions or 'democratising' heritage research by making accurate documentation or models available to wider groups of researchers or amateurs. However, technology also opens up an avenue of nationalisms by making it easier to reconstruct or interpret erroneously in a digital world. It also influences the tourism market. The pre-tour sales process is not only a description of intangible experiences anymore. The author's observations at the World Travel Market events in the last few years witnessed a strong trend towards the use of technology, from simple videos to virtual reality headsets, to promote destinations of archaeological or heritage tourism. It is becoming popular to analyse the role of social media within heritage tourism (Park, 2013) and the impact that it has on the increase in visitor numbers (Miller, 2017) or on the overcrowding of World Heritage Sites (Fletcher *et al.*, 2017; Frary, 2017). Youth travel is one of the fastest growing segments and currently represents more than 23% of travellers globally (UNWTO, 2016a: 10) and, in addition to higher awareness of technology, they are also bringing different spending and destination selection implications to the sector (ITB Berlin, 2016).

Further discussions on the interaction of the three key concepts in this chapter, particularly in light of the changing understanding of the benefits of archaeological tourism and shifting agendas towards longer-term sustainability, are of course necessary and important. Theoretical discussions around heritage and tourism are gaining new momentum (Staiff *et al.*, 2013). Archaeology can contribute to society in many ways in addition to tourism, but archaeology as a discipline may need to rethink how it makes the knowledge it possesses available for public consumption and public policy (Darvill, 2015). There are some major issues in understanding the impacts of archaeological or heritage tourism in various localities and increased availability of clear statistics on visitor numbers at many major archaeological sites, especially in developing world, may shed more light on potential trends for the future.

References

Adler, J. (1989) Travel as performed art. *American Journal of Sociology* 94 (6), 1366–1391.
AIA (2017) A Guide to Best Practices for Archaeological Tourism, Archaeological Institute of America. See https://www.archaeological.org/tourism_guidelines (accessed 3 December 2017).
Aitchison, K. (2009) After the 'Gold Rush': Global archaeology in 2009. *World Archaeology* 41 (4), 659–671.
Aitchison, K. (2015) Professional archaeology in the UK in 2015. *Cultural Trends* 24 (1), 11–14.
Anderson, B.R.O. (1983) *Imagined Communities: Reflections on the Origin and Spread of Nationalism*. London: Verso.
Anderson, W. (2015) Cultural tourism and poverty alleviation in rural Kilimanjaro, Tanzania. *Journal of Tourism and Cultural Change* 13 (3), 208–224.
Anglin, R. (2008) The World Heritage List: Bridging the cultural property nationalism-internationalism divide notes. *Yale Journal of Law & the Humanities* 20, 241–276.
Arts Council England (2012) *Measuring the Economic Benefits of Arts and Culture*. Manchester: Arts Council England.
Arts Council England (2013) *Great Art and Culture for Everyone: 10 Year Strategic Framework, 2010–2020* (2nd edn) Manchester: Arts Council England.
Arts Council England (2014) *The Value of Arts and Culture to People and Society – An Evidence Review*. Manchester: Arts Council England.
Askew, M. (2010) The magic list of global status: UNESCO, World Heritage and the agendas of states. In S. Labadi and C. Long (eds) *Heritage and Globalisation* (pp. 19–44). London: Routledge.
Atkinson, J.A., Banks, I. and O'Sullivan, J. (eds) (1996) *Nationalism and Archaeology: Scottish Archaeological Forum*. Glasgow: Cruithne Press.
Bandarin, F., Hosagrahar, J. and Albernaz, F.S. (2011) Why development needs culture. *Journal of Cultural Heritage Management and Sustainable Development* 1(1), 15–25.
Beaumont, P. (2017) UNESCO makes Hebron old city Palestinian World Heritage Site. *The Guardian*. See http://www.theguardian.com/world/2017/jul/07/unesco-recognises-hebron-as-palestinian-world-heritage-site (accessed 4 April 2018).
Bender, A. (2013) Two-Thirds Of Travelers Want Green Hotels: Here's How To Book Them. *Forbes*. See https://www.forbes.com/sites/andrewbender/2013/04/22/survey-two-thirds-of-travelers-want-green-hotels-heres-how-to-book-them/ (accessed 28 December 2017).
Burtenshaw, P. (2014) Mind the gap: Cultural and economic values in archaeology. *Public Archaeology* 13 (1–3), 48–58.
Burtenshaw, P. (2017) Economics in public archaeology. In G. Moshenska (ed.) *Key Concepts in Public Archaeology* (pp. 31–42). London: UCL Press

Burtenshaw, P. and Palmer, C. (2014) Archaeology, local development and tourism: A role for international institutes. *Bulletin for the Council for British Research in the Levant* 9(1), 21–26.

Butcher, J. (2003) *The Moralisation of Tourism: Sun, Sand... and Saving the World?* London: Routledge.

Butler, R.W. (1999) Sustainable tourism: A state-of-the-art review. *Tourism Geographies* 1(1), 7–25.

Carman, J. and Sørensen, M.L.S. (2009) Heritage studies: An outline. In M.L.S. Sørensen and J. Carman (eds) *Heritage Studies: Methods and Approaches* (pp. 11–28). London: Routledge.

CEBR (2013) *The Contribution of the Arts and Culture to the National Economy.* See http://www.artscouncil.org.uk/sites/default/files/download-file/The_contribution_of_the_arts_and_culture_to_the_national_economy.pdf (accessed 4 January 2018).

CEBR (2015) *Contribution of the Arts and Culture Industry to the National Economy.* See http://www.artscouncil.org.uk/sites/default/files/download-file/Arts_culture_contribution_to_economy_report_July_2015.pdf (accessed 4 January 2018).

Cellini, R. (2011) Is UNESCO recognition effective in fostering tourism? A comment on Yang, Lin and Han. *Tourism Management* 32(2), 452–454.

Cernea, M.M. (2001) *Cultural Heritage and Development: A Framework for Action in the Middle East and North Africa.* Washington, DC: The World Bank.

Champion, T.C. and Díaz-Andreu, M. (eds) (1996) *Nationalism and Archaeology in Europe.* London: UCL Press.

Coningham, R. (2017) Why the US withdrawal from UNESCO is a step backwards for global cultural cooperation. *The Conversation.* See http://theconversation.com/why-the-us-withdrawal-from-unesco-is-a-step-backwards-for-global-cultural-cooperation-85692 (accessed 4 April 2018).

Connolly, K. (2017) Cruise tourists overwhelm Europe's ancient resorts. *BBC News.* See http://www.bbc.co.uk/news/world-europe-40592247 (accessed 3 December 2017).

Connor, W. (1993) *Ethnonationalism: The Quest for Understanding.* Princeton, NJ: Princeton University Press.

Corbishley, M. (2011) *Pinning Down the Past: Archaeology, Heritage, and Education Today.* Woodbridge: Boydell.

Corsane, G. (ed.) (2005) *Heritage, Museums and Galleries: An Introductory Reader.* London: Routledge.

Council of Europe (2005) Council of Europe Framework Convention on the Value of Cultural Heritage for Society. See http://conventions.coe.int/Treaty/EN/Treaties/Html/199.htm (accessed 1 September 2014).

Cuccia, T. and Rizzo, I. (2013) Seasonal tourism flows in UNESCO sites: The case of Sicily. In J. Kaminski, A.M. Benson and D. Arnold (eds) *Contemporary Issues in Cultural Heritage Tourism* (pp. 179–199). London: Routledge

Darvill, T. (2015) Making futures from the remains of the distant past. In M.H. van den Dries, S.J. van der Linde and A. Strecker (eds) *Fernweh: Crossing Borders and Connecting People in Archaeological Heritage Management* (pp. 42–46). Leiden, Netherlands: Sidestone Press

DCMS (2015) Culture and Creativity: Yesterday, Today and Tomorrow. See http://www.thecreativeindustries.co.uk/media/290639/culture-and-creativity-yesterday-today-and-tomorrow-2-.pdf (accessed 3 April 2018).

Díaz-Andreu García, M. (2007) *A World History of Nineteenth-century Archaeology.* Oxford: Oxford University Press.

Díaz-Andreu García, M. and Champion, T.C. (eds) (2015) *Nationalism and Archaeology in Europe.* London: Routledge.

Dragouni, M. (2017) Sustainable heritage tourism: Towards a community-led approach. UCL University College London. Unpublished doctoral thesis, University College of London.

Edensor, T. (2001) Performing tourism, staging tourism: (Re)producing tourist space and practice. *Tourist Studies* 1 (1), 59–81.

Eriksen, T.H. (1993) *Ethnicity and Nationalism: Anthropological Perspectives*. London: Pluto Press.

European Commission (2018) Supporting cultural and creative industries – Culture. See / culture/policy/cultural-creative-industries_en (accessed 6 April 2018).

Flatman, J. (2012) Conclusion: The contemporary relevance of archaeology – Archaeology and the Real World? In M. Rockman and J. Flatman (eds) *Archaeology in Society: Its Relevance in the Modern World* (pp. 291–303). London: Springer

Fletcher, J., Fyall, A., Gilbert, D. and Wanhill, S. (2017) *Tourism: Principles and Practice* (6th edn). London: Pearson.

Fowler, D.D. (1987) Uses of the past: Archaeology in the service of the state. *American Antiquity* 52(2), 229–248.

Frary, M. (2017) World Heritage: Spreading the love. *Travel Perspective*. See http://www.travelperspective.co.uk/2017/11/20/world-heritage-spreading-the-love/ (accessed 4 April 2018).

Frey, B.S. and Steiner, L. (2011) World Heritage List: Does it make sense? *International Journal of Cultural Policy* 17 (5), 555–573.

Galaty, M.L. and Watkinson, C. (2004) The practice of archaeology under dictatorship. In M.L. Galaty and C. Watkinson (eds) *Archaeology under Dictatorship* (pp. 1–18). London: Kluwer Academic/Plenum Publishers.

Gellner, E. (1983) *Nations and Nationalism*. Oxford: Blackwell.

Gellner, E. (2008) *Nations and Nationalism*. Ithaca, NY: Cornell University Press.

Gestrich, N. (2011) Putting a price on the past: The ethics and economics of archaeology in the marketplace – A reply to 'What is Public Archaeology'. *Present Pasts* 3 (2), n.p. (online journal)

Girard, L.F. and Nijkamp, P. (eds) (2009) *Cultural Tourism and Sustainable Local Development*. Farnham: Ashgate.

Graham, B., Ashworth, G.J. and Tunbridge, J.E. (2005) The uses and abuses of heritage. In G. Corsane (ed.) *Heritage, Museums and Galleries: An Introductory Reader* (pp. 28–40). London: Routledge.

Guibernau, M. (2004) Anthony D. Smith on nations and national identity: A critical assessment. *Nations and Nationalism* 10 (1–2), 125–141.

Hall, C.M. and Piggin, R. (2003) World Heritage Sites: Managing the brand. In A. Fyall, B. Garrod and A. Leask (eds) *Managing Visitor Attractions: New Directions* (pp. 203–220). Oxford: Butterworth-Heinemann

Hall, D. and Richards, G. (eds) (2000) *Tourism and Sustainable Community Development*. London: Routledge.

Harrison, R. (2013) *Heritage: Critical Approaches*. London: Routledge.

Hems, A. and Blockley, M. (eds) (2006) *Heritage Interpretation: Theory and Practice*. London: Routledge.

Herbert, D. (ed.) (1995) *Heritage, Tourism and Society*. London: Mansell.

Hobsbawm, E.J. and Ranger, T.O. (eds) (1983) *The Invention of Tradition*. Cambridge: Cambridge University Press.

Hughes, R. (2008) Dutiful tourism: Encountering the Cambodian genocide. *Asia Pacific viewpoint* 49 (3), 318–330.

Hutchinson, J. and Smith, A.D. (eds) (1994) *Nationalism*. Oxford: Oxford University Press.

Hutchinson, J. and Smith, A.D. (eds) (1996) *Ethnicity*. Oxford: Oxford University Press.

ICOMOS (1999) ICOMOS International Cultural Tourism Charter. Managing Tourism at Places of Heritage Significance. See https://www.icomos.org/charters/tourism_e.pdf (accessed 4 January 2018).

ICOMOS and UNWTO (1996) *Tourism at World Heritage Cultural Sites: The Site Manager's Handbook* (2nd edn). Madrid: World Tourism Organization.

ITB Berlin (2015) *ITB World Travel Trends Report 2015/2016*. Berlin: Messe Berlin GmbH.
ITB Berlin (2016) *ITB World Travel Trends Report 2016/2017*. Berlin: Messe Berlin GmbH.
ITB Berlin (2018) *ITB World Travel Trends Report 2017/2018*. Berlin: Messe Berlin GmbH.
Jimura, T. (2011) The impact of World Heritage Site designation on local communities: A case study of Ogimachi, Shirakawa-mura, Japan. *Tourism Management* 32 (2), 288–296.
Junker, K. (1998) Research under dictatorship: The German Archaeological Institute 1929-1945. *Antiquity* 72 (276), 282.
Kaminski, J., Benson, A.M. and Arnold, D. (eds) (2013) *Contemporary Issues in Cultural Heritage Tourism*. London: Routledge.
Kaufmann, E. (2017) Complexity and nationalism. *Nations and Nationalism* 23 (1), 6–25.
Kettle, M. (2017) Mass tourism is at a tipping point – but we're all part of the problem. *The Guardian*. See http://www.theguardian.com/commentisfree/2017/aug/11/tourism-tipping-point-travel-less-damage-destruction (accessed 3 December 2017).
Kim, S.S., Timothy, D.J. and Han, H.-C. (2007) Tourism and political ideologies: A case of tourism in North Korea. *Tourism Management* 28 (4), 1031–1043.
Kirshenblatt-Gimblett, B. (1998) *Destination Culture: Tourism, Museums and Heritage*. Berkeley, CA: University of California Press.
Klamer, A. (2014) The values of archaeological and heritage sites. *Public Archaeology* 13 (1–3), 59–70.
KMU Forschung Austria and VVA Europe (2016) Boosting the competitiveness of cultural and creative industries for growth and jobs. See /growth/content/boosting-competitiveness-cultural-and-creative-industries-growth-and-jobs-0_en (accessed 6 April 2018).
Knapp, A.B. and Antoniadou, S. (1998) Archaeology, politics and the cultural heritage of Cyprus. In L. Meskell (ed.) *Archaeology under Fire: Nationalism, Politics and Heritage in the Eastern Mediterranean and Middle East* (pp. 13–43). London: Routledge
Kohl, P.L. (1998) Nationalism and archaeology: On the constructions of nations and the reconstructions of the remote past. *Annual Review of Anthropology* 27, 223–246.
Kohl, P.L. and Fawcett, C. (1996) *Nationalism, Politics and the Practice of Archaeology*. Cambridge: Cambridge University Press.
Kohl, P.L., Kozelsky, M. and Ben-Yehuda, N. (eds) (2007) *Selective Remembrances: Archaeology in the Construction, Commemoration, and Consecration of National Pasts*. Chicago: University of Chicago Press.
Labadi, S. and Long, C. (2010) Introduction. In S. Labadi and C. Long (eds) *Heritage and Globalisation* (pp. 1–16). London: Routledge
Lange, L. (2011) *Exploring the Leakage Effect in Tourism in Developing Countries – Issues and Implications*. Bonn: International University of Applied Sciences Bad Honnef – BONN.
Larkin, J. (2014) Safely into the unknown? A review of the proposals for the future of English Heritage. *Papers from the Institute of Archaeology* 24 (1), n.p (online journal)
Lee, S., Phau, I. and Quintal, V. (2018) Exploring the effects of a 'new' listing of a UNESCO World Heritage Site: The case of Singapore Botanic Gardens. *Journal of Heritage Tourism* 13 (4), 339–355.
Lehnes, P. and Carter, J. (2016) *Digging Deeper: Exploring the Philosophical Roots of Heritage Interpretation*. Waldkirch: InHerit.
Lennox, R. (2018) Landward Research publishes new report into archaeological sector market demand. See http://archaeologists.net/news/landward-research-publishes-new-report-archaeological-sector-market-demand-1515585049 (accessed 21 January 2018).
Lynch, C. (2017) U.S. to pull out of UNESCO, again. *Foreign Policy*. See https://foreignpolicy.com/2017/10/11/u-s-to-pull-out-of-unesco-again/ (accessed 4 April 2018).

Lyon, S. and Wells, E.C. (eds) (2012) *Global Tourism: Cultural Heritage and Economic Encounters*. Lanham, MD: AltaMira Press.

Macdonald, S. (ed.) (2010) *A Companion to Museum Studies*. Chichester: Wiley.

Maurel, C. (2017) The unintended consequences of UNESCO World Heritage listing. *The Conversation*. See http://theconversation.com/the-unintended-consequences-of-unesco-world-heritage-listing-71047 (accessed 4 April 2018).

Mergos, G. and Patsavos, N. (eds) (2017) *Cultural Heritage and Sustainable Development: Economic Benefits, Social Opportunities and Policy Challenges*. Chania: Technical University of Crete.

Meskell, L. (ed.) (1998) *Archaeology under Fire: Nationalism, Politics and Heritage in the Eastern Mediterranean and Middle East*. London: Routledge.

Meskell, L. (2002) The intersections of identity and politics in archaeology. *Annual Review of Anthropology* 31 (1), 279–301.

Meskell, L. (2015) Transacting UNESCO World Heritage: Gifts and exchanges on a global stage. *Social Anthropology* 23 (1), 3–21.

Miller, C. (2017) How Instagram is changing travel. *National Geographic*. See https://www.nationalgeographic.com/travel/travel-interests/arts-and-culture/how-instagram-is-changing-travel/ (accessed 17 December 2017).

Moser, S. (2010) The devil is in the detail: Museum displays and the creation of knowledge. *Museum Anthropology* 33 (1), 22–32.

Moussouri, T. (2014) From 'telling' to 'consulting': A perspective on museums and modes of public engagement. In S. Thomas and J. Lea (eds) *Public Participation in Archaeology* (pp. 119–128). Woodbridge: Boydell.

Mowforth, M. and Munt, I. (2016) *Tourism and Sustainability: Development, Globalisation and New Tourism in the Third World* (4th edn). London: Routledge.

Newell, G.E. (2008) Rhyming culture, heritage, and identity: The 'total site' of Teotihuacan, Mexico. *Journal of Heritage Tourism* 3 (4), 243-255.

Nickerson, N.P., Jorgenson, J. and Boley, B.B. (2016) Are sustainable tourists a higher spending market? *Tourism Management* 54, 170–177.

Norman, K. (2011) Should the UK be nominating more World Heritage Sites? *Present Pasts* 3 (2), n.p. (online journal)

O'Brien, D. (2010) *Measuring the Value of Culture: A Report to the Department for Culture Media and Sport*. See https://assets.publishing.service.gov.uk/government/uploads/system/uploads/attachment_data/file/77933/measuring-the-value-culture-report.pdf (accessed 4 January 2018).

O'Leary, B. (1997) On the nature of nationalism: An appraisal of Ernest Gellner's writings on nationalism. *British Journal of Political Science* 27 (2), 191–222.

Orbaşli, A. (2013) Archaeological site management and local development. *Conservation and Management of Archaeological Sites* 15 (3–4), 237–253.

Park, H.Y. (2013) *Heritage Tourism*. London: Routledge.

Pawakapan, P.R. (2013) *State and Uncivil Society in Thailand at the Temple of Preah Vihear*. Singapore: Institute of Southeast Asian Studies.

Pearce, S.M. (1994) *Interpreting Objects and Collections Edited by Susan M. Pearce*. London: Routledge.

Pedersen, A. (2002) *Managing Tourism at World Heritage Sites: A Practical Manual for World Heritage Site Managers*. Paris: UNESCO World Heritage Centre.

Peter, L. (2017) 'Tourists go home': Leftists resist Spain's influx. *BBC News*. See http://www.bbc.co.uk/news/world-europe-40826257 (accessed 3 December 2017).

Pulido-Fernández, J.I. and López-Sánchez, Y. (2016) Are tourists really willing to pay more for sustainable destinations? *Sustainability* 8 (12), 1–20.

Rasoolimanesh, S.M. and Jaafar, M. (2017) Sustainable tourism development and residents' perceptions in World Heritage Site destinations. *Asia Pacific Journal of Tourism Research* 22 (1), 34–48.

Rebanks Consulting (2009) *UNESCO World Heritage Site status: Is there opportunity for economic gain?* See http://icomos.fa.utl.pt/documentos/2009/WHSTheEconomic GainFinalReport.pdf (accessed 10 December 2017).

Rifai, T. (2017) Tourism: Growth is not the enemy; it's how we manage it that counts. *World Tourism Organization UNWTO.* See http://media.unwto.org/press-release/2017-08-15/tourism-growth-not-enemy-it-s-how-we-manage-it-counts (accessed 3 April 2018).

Roberts, L.C. (1997) *From Knowledge to Narrative: Educators and the Changing Museum.* Washington, DC: Smithsonian Books.

Robinson, M. and Picard, D. (2006) *Tourism, Culture and Sustainable Development.* Paris: UNESCO.

Rock-Macqueen, D. (2018) *Have We Reached Peak Archaeologists (in the UK)? A Landward Research White Paper on Commercial Archaeology Job Demand.* Sheffield: Landward Research Ltd.

Roe, D. and Urquhart, P. (2002) *Pro-Poor Tourism: Harnessing the World's Largest Industry for the World's Poor.* London: *International Institute for Environment and Development.*

Salazar, N.B. and Zhu, Y. (2015) Heritage and tourism. In L. Meskell (ed.) *Global Heritage: A Reader* (pp. 240–258). Chichester: Wiley Blackwell

Schech, S. and Haggis, J. (2000) *Culture and Development: A Critical Introduction.* Chichester: Wiley.

Shanks, M. and Tilley, C.Y. (1992) *Re-constructing Archaeology: Theory and Practice.* Cambridge: Cambridge University Press.

Shnirelman, V. (2009) Presidents and archeology, or what do politicians seek in ancient times: Distant past and its political role in the USSR and during the post-Soviet period (Президенты и археология, или что ищут политики в древности: далекое прошлое и его политическая роль в СССР и в постсоветское время). *Ab Imperio* 1/2009, 279–324.

Smith, A.D. (1991) *The Ethnic Origins of Nations.* Chichester: Wiley.

Smith, L. (2006) *The Uses of Heritage.* London: Routledge.

Smith, M.K. (2003) *Issues in Cultural Tourism Studies.* London: Routledge.

Sommer, U. (2017) Archaeology and nationalism. In G. Moshenska (ed.) *Key Concepts in Public Archaeology* (pp. 166–186). London: UCL Press.

Sørensen, M.L.S. (1996) The fall of a nation, the birth of a subject: The national use of archaeology in nineteenth-century Denmark. In M. Díaz-Andreu and T. Champion (eds) *Nationalism and Archaeology in Europe* (pp. 24–47). London: UCL Press.

Staiff, R., Bushell, R. and Watson, S. (eds) (2013) *Heritage and Tourism: Place, Encounter, Engagement.* London: Routledge.

Su, M.M. and Wall, G. (2012) Community participation in tourism at a World Heritage Site: Mutianyu Great Wall, Beijing, China. *International Journal of Tourism Research* 16 (2), 146–156.

Suny, R.G. (2001) Constructing primordialism: Old histories for new nations. *The Journal of Modern History* 73 (4), 862–896.

Thomas, S. and Lea, J. (eds) (2014) *Public Participation in Archaeology.* Woodbridge: Boydell Press.

Tilden, F. (1957) *Interpreting Our Heritage: Principles and Practices for Visitor Services in Parks, Museums, and Historic Places.* Chapel Hill, NC: University of North Carolina Press.

Timothy, D.J. (2011) *Cultural Heritage and Tourism: An Introduction.* Bristol: Channel View Publications.

Timothy, D.J. (2014) Contemporary cultural heritage and tourism: Development issues and emerging trends. *Public Archaeology* 13 (1–3), 30–47.

Timothy, D.J. and Boyd, S.W. (2003) *Heritage Tourism.* London: Prentice Hall.

Timothy, D.J. and Boyd, S.W. (2006) Heritage tourism in the 21st century: Valued traditions and new perspectives. *Journal of Heritage Tourism* 1 (1), 1–16.
Treanor, P. (1997) Structures of Nationalism. *Sociological Research Online* 2(1), n.p. (online journal) See http://www.socresonline.org.uk/2/1/8.html (accessed 24 November 2011).
Trigger, B.G. (1984) Alternative archaeologies: Nationalist, colonialist, imperialist. *Man* 19 (3), 355–370.
UN (2018) Sustainable development knowledge platform. See https://sustainabledevelopment.un.org (accessed 4 April 2018).
UNCTAD (2013) Sustainable tourism: Contribution to economic growth and sustainable development. See http://unctad.org/meetings/en/SessionalDocuments/ciem5d2_en.pdf (accessed 3 January 2018).
UNESCO (2017a) Culture: a bridge to development. *United Nations Educational, Scientific and Cultural Organization*. See http://www.unesco.org/new/en/venice/culture/culture-a-bridge-to-development/ (accessed 18 February 2018).
UNESCO (2017b) Culture and development. *United Nations Educational, Scientific and Cultural Organization*. See http://www.unesco.org/new/en/culture/themes/culture-and-development/ (accessed 18 February 2018).
UNESCO WHC (2011) Recommendation on the historic urban landscape. See http://whc.unesco.org/en/hul/ (accessed 18 February 2018).
United Nations (2016) International Year of Sustainable Tourism for Development, 2017. Resolution adopted by the General Assembly. See https://undocs.org/A/RES/70/193 (accessed 19 December 2017).
UNWTO (1997) *Agenda 21 for the Travel and Tourism Industry: Towards Environmentally Sustainable Development*. Madrid: World Tourism Organization.
UNWTO (2004) *Tourism at World Heritage Cultural Sites*. Madrid: World Tourism Organization.
UNWTO (2016a) *Global Report on the Power of Youth Travel*. Madrid: UNWTO.
UNWTO (2016b) Silk Road Action Plan 2016/2017. See http://cf.cdn.unwto.org/sites/all/files/docpdf/sr2016web.pdf (accessed 19 November 2017).
UNWTO (2017a) *UNWTO Tourism Highlights: 2017 Edition*. See https://www.e-unwto.org/doi/book/10.18111/9789284419029 (accessed 3 December 2017).
UNWTO (2017b) *UNWTO World Tourism Barometer, Volume 15*. See http://cf.cdn.unwto.org/sites/all/files/pdf/unwto_barom17_06_december_excerpt_.pdf (accessed 3 April 2018).
UNWTO (2018) 2017 International tourism results: the highest in seven years. *World Tourism Organization UNWTO*. See http://media.unwto.org/press-release/2018-01-15/2017-international-tourism-results-highest-seven-years (accessed 3 April 2018).
Urry, J. and Larsen, J. (2011) *The Tourist Gaze 3.0*. London: Sage.
VanBlarcom, B.L. and Kayahan, C. (2011) Assessing the economic impact of a UNESCO World Heritage designation. *Journal of Heritage Tourism* 6 (2), 143–164.
Wang, Y., Broeck, A.M.V. and Vanneste, D. (2017) International tourism in North Korea: How, where and when does political ideology enter? *International Journal of Tourism Cities* 3 (3), 260–272.
Weaver, D.B. (2012) Organic, incremental and induced paths to sustainable mass tourism convergence. *Tourism Management* 33 (5), 1030–1037.
Willems, A. and Dunning, C. (2015) Solving the puzzle: The characteristics of archaeological tourism. In M.H. van den Dries, S.J. van der Linde and A. Strecker (eds) *Fernweh: Crossing Borders and Connecting People in Archaeological Heritage Management* (pp. 68–71). Leiden, Netherlands: Sidestone Press
Williams, T. (2011) The curious tale of Preah Vihear: The process and value of World Heritage nomination. *Conservation and Management of Archaeological Sites* 13 (1), 1–7.

Winter, T. (2010) Heritage tourism: The dawn of a new era? In S. Labadi and C. Long (eds) *Heritage and Globalisation* (pp. 117–129). London: Routledge

Winter, T. (2012) Heritage and nationalism: An unbreachable couple? *Institute for Culture and Society Occasional Paper* 3 (4), 1–13.

Winter, T. (2015) Heritage and nationalism: An unbreachable couple? In E. Waterton and S. Watson (eds) *The Palgrave Handbook of Contemporary Heritage Research* (pp. 331–345). London: Palgrave Macmillan.

Wogan, P. (2001) Imagined communities reconsidered: Is print-capitalism what we think it is? *Anthropological Theory* 1 (4), 403–418.

Wood, M.E. (2017) *Sustainable Tourism on a Finite Planet: Environmental, Business and Policy Solutions*. London: Routledge.

World Heritage Centre (2011) UNESCO Special Envoy on Preah Vihear to meet with prime ministers of Thailand and Cambodia. *UNESCO World Heritage Centre*. See http://whc.unesco.org/en/news/715/ (accessed 28 December 2017).

13 Understanding Perspectives on Archaeology and Tourism

Dallen J. Timothy and Lina G. Tahan

Readers of this volume will by now be aware that archaeology is one of the most ubiquitous tourism resources in the world and one of the best-known manifestations of cultural heritage. Archaeological attractions include historic monuments, ruins, active excavations, museums, and interpretive centers, to name only a few. People visit archaeological attractions for many reasons, including self-edification, to gain knowledge and understanding, pursue a hobby, learn a skillset, have a leisure experience, socialize with others, or become immersed in a cultural landscape that is different from the one where they live. The supply of, and demand for, archaeological sites resembles those of general heritage tourism. However, they oftentimes have their own characteristics, management challenges, advantages and disadvantages.

Several important themes arose from the essays herein that deserve more emphasis. This chapter examines some of these in greater detail, after which it raises questions that are worthwhile addressing through additional research and scholarly inquiry. Issues raised in the book include heritage branding and politics, scale and spatial variability, archaeology-based niche tourisms, the relationships between archaeology and different types of tourism, overtourism and its damaging implications and archaeology and indigenous people. Other important issues to consider for future inquiry include climate change and information communication technology and their relationships with archaeology-based tourism.

Branding Archaeological Heritage and Politics

Some contributions to this volume have addressed the notion of the utilization of archaeology for heritage branding. For the tourism enterprise, branding is an important element of promotion and can help stimulate growth and offer a unique selling proposition for destinations that compete one with another. Archaeological heritage may provide the advantage needed to be competitive in the tourism marketplace. Several countries are extremely eager to inscribe their heritage locales on the UNESCO World Heritage List under the nebulous assumption that this will inevitably result in increased numbers of tourists (Ribaudo & Figini, 2017). China and Italy are particularly enthusiastic about this prospect and continue to have the highest numbers of WHSs (Figure 13.1). World Heritage is the most globally recognized heritage brand, although several scholars have argued that this coveted trademark is frequently more about favoritism and politicking than it is about the merits of 'universal value' (Meskell, 2013; Meskell *et al.*, 2015; Schmitt, 2009; Scholze, 2008). In Logan's (2013) opinion, it is the states parties that are the biggest culprits in politicizing the World Heritage Convention for their own national interests, not the UNESCO organization itself.

While research about heritage branding is growing, we still know relatively little about it. The WHS brand is believed by many to increase tourist arrivals automatically, but research results are mixed, and this depends on individual contexts. In locations that are accessible, inexpensive, secure, and prominent, it is highly likely that the added visibility of UNESCO might in fact increase visitor arrivals. However, in isolated contexts, in archaeological places that are not already popular attractions, the added brand will probably not have the same effect. While many nationalities are aware of the WHS brand, many others are not. In some countries, the brand has little effect and people are unaware of what it means

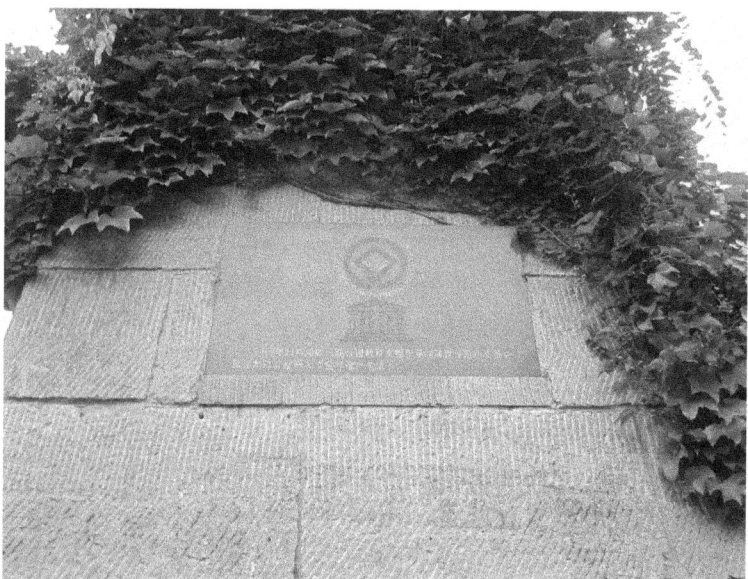

Figure 13.1 Many countries, including China, are especially eager to extend the UNESCO 'brand' to as many archaeological sites as possible (Photo: Dallen J. Timothy)

(Boyd & Timothy, 2006). These conditions are important to understand in context-specific situations. WHSs in the United States and India are treated very differently among domestic tourists than they are in China, for example. While most Chinese travelers are aware of the importance of the UNESCO brand and see WHSs as national treasures, most Americans are entirely unaware of what the brand means (Boyd & Timothy, 2006). Likewise, during various trips to India, both authors of this chapter have noticed a general unawareness of the meaning and value of WHSs there. Many Indian WHSs are abused by tourists because of poor policy enforcement. Sometimes, efforts to maintain or restore WHSs cause more damage than leaving them alone, such as so-called restorers painting over original Victorian tiles in the Chhatrapati Shivaji Maharaj (Victoria Terminus) train station in Mumbai.

Scale and Geographical Variability

Scale and spatial distribution are important considerations in understanding archaeology-based tourism. The archaeological record is everywhere. Every country has a human history, although some countries possess larger numbers of artifacts than others do. For instance, some states have massive volumes of archaeological remains, even to the extent that many layers of archaeological strata exist, telling diverse stories of national development. Many countries are home to plentiful buried and exposed remains that tell a multiplicity of stories about human history and provide a valuable asset for tourism development. The United Kingdom, India, Egypt, Turkey, Greece, Italy, Spain, Mexico, Peru, China and Cambodia are only a handful of countries with a famed archaeological record. Some of these and other countries are so full of archaeology that communities, regions and specific sites must wrangle for funding and staff support in an increasingly competitive heritage and tourism environment. Thus, difficult decisions have to be made regarding which heritage should be, or can be, excavated, preserved and interpreted. In many developing countries, this is a significant problem where personnel and budget shortages limit the scientific and protective work that can be done, because there are simply too many archaeological remains to protect.

From a tourism perspective, this results in certain archaeologies receiving much larger tourism attention than others so that 'the supply of cultural heritage sites has outgrown demand' (Berry & Shephard, 2001: 159). This creates a problem in that most tourism (and therefore economic) attention has long been geared toward the extraordinary built heritage and most lavish monuments at the expense of the vernacular heritage of ordinary people and places (Timothy, 2014a, 2014b). Thus, attention to the Chichen Itzas, Teotihuacans and Machu Picchus of the world has far surpassed the attention given to the vernacular heritagescapes that surround them, even though these everyday heritagescapes are likely more connected to the personal cultural lives of the descendant communities that inhabit them. Heritage tourism's values have long been assigned to the historical provenance of kings, queens, priests and national heroes, yet these represent an infinitesimal proportion of people who have heretofore dwelt on the Earth (Timothy, 2018). Questions must address how ordinary heritages, cultural landscapes and archaeologies are consumed or why they are not, and how changing this pattern might help empower communities and distribute the public demand for archaeology more equitably.

Geographical scale relates to the point above but can be applied to the size of nations. The archaeological record in the smallest states typically goes unnoticed by tourists, even though for these countries they are a significant foundation of their national identities, societal nostalgia and ethnic solidarity. For example, the ruins of Obere Burg and Untere Burg castles in Liechtenstein are rarely visited by tourists, who prefer to spend time shopping in the capital, Vaduz, or at more 'important' historic localities in nearby Switzerland or Austria. Yet, these two archaeological remains are extremely important heritage sites in this microstate that wield considerable national pride for Liechtensteiners. Likewise, despite its incredible collection of Romanesque churches and bridges, Andorra's tourism focus has long been shopping and skiing, despite its much greater potential for heritage tourism. While the country does actively promote its heritage assets as part of its tourism product, the size of the country and tourists' focus on Andorra's skiing and shopping sectors means that much of its heritage goes relatively unnoticed.

Spatial variations in settlement types and patterns also means an unequal distribution of certain archaeological remains. In some locations, stones or clay were used to build dwellings, while in other localities, shelters were made from wood, fronds, or animal skins. This was determined largely by what construction materials were available, but also by cultural traditions and method of subsistence (e.g. itinerant hunting or settled farming). Thus, the remaining record depends on the durability or resilience of the materials used and the permanence of the structures. Where nomadic hunting-and-gathering societies dominated rather than permanent agricultural settlements, there are fewer built structures to excavate. Nomadic peoples left behind tools, bones, fire pits and rock art but few buildings. In some Pacific Islands, in Aboriginal Australia, and in some ancient American tribal areas, archaeological remains are different from those where large, permanent communities existed. The remains of permanent settlements may tell broader stories and serve as keepers of more data than the scattered artifacts and encampments of nomadic peoples, yet both are crucial in understanding the archaeological record.

Most tourism focuses on the most durable archaeological artifacts – buildings and ruins – because these are what remain most intact and in their original locations, and they tend to be more momentous and marketable. Nevertheless, much of the archaeological record of people who did not live in permanent settlements has become an important attraction as well. Their daily artifacts (e.g. hunting tools and pottery) feature in many museums across the world and contribute much to our understanding of ancient life. Likewise, rock art appears to be one of the most abiding visitor favorites in places where there are few other permanent artifacts and records (Deacon, 2006; Duval & Smith, 2013; Wurtz & van der Merwe, 2005), although these are also found in areas where native people lived in stationary settlements.

A final geographical perspective is the recent trend in places that have traditionally relied on one tourism type, most especially sun, sea and sand (SSS) tourism, diversifying their products to include more heritage (Cameron & Gatewood, 2008; Jordan & Jolliffe, 2013). While the Caribbean has a rich and diverse cultural heritage, and a few countries (e.g. Jamaica, Haiti and the Dominican Republic) have utilized heritage as a tourism asset, many of the region's islands have long focused solely on the SSS product. Several countries, however, including Barbados, Jamaica, the Bahamas, and the Turks and Caicos Islands have now begun to tap into their 'hidden' heritage related to slavery and sugar, and New World 'discovery' to supplement the increasingly competitive beach-based leisure market (Cameron & Gatewood, 2008; Jolliffe, 2013). Archaeology is playing an increasingly important role in this as obscured sites and ruins of slave quarters, plantations and cemeteries are finally receiving scholarly attention commensurate with their important role in the development of the Caribbean states (e.g. Catalani & Ackroyd, 2013). Future efforts should be geared towards realizing the role of archaeology in diversifying the tourism product in traditional mass tourism destinations. What ancillary appeal does it add? In what ways can archaeological heritage enhance the attractiveness of destinations that are associated with non-heritage forms of tourism? How willing are tourists to spend significant time during their leisure holidays visiting archaeological sites that help establish a place's sense of identity?

Elements of Material Culture – Niches in Archaeology-based Tourism

Archaeologists unearth and discover many different types of material culture: tools, bones, hunting accoutrements and projectile points, fire pits, building materials, pottery, glass, toys, coins, jewelry and many other artifacts. From a tourism perspective, different artifacts might appeal to different audiences. As noted previously, some people visit archaeological areas based on their hobby interests, and serious heritage tourists will make the effort to visit certain sites to learn about topics of interest. A numismatist (coin collector), for example, might be particularly interested in seeing a hoard of silver pieces unearthed at a Viking site, ancient Greek coins at an archaeology museum in Thessaloniki, or wampum at a Native American museum (Figure 13.2).

People with a particular interest in Roman heritage would be keen to visit Hadrian's Wall, the Roman Forum in Rome, the Temple of Bacchus in Baalbek, Lebanon, or sections of the Eifel Aqueduct in Germany. Viking hobbyists may have written the Jorvik Viking Centre in York, England, the ruins of the Hvalsey settlement near Qaqortoq, Greenland, and the L'Anse aux Meadows National Historic Site in Newfoundland, Canada, on their list of must-see places to visit. For biblical enthusiasts, there are hundreds of sites to visit in Egypt, Palestine, Israel, Jordan, Lebanon, Syria, Iraq, Turkey, Greece, Italy and Cyprus associated with the Old Testament, the New Testament, the life of Christ, and the spreading of the Christian message. Silk Road hobbyists have many sites to choose from in China, Turkmenistan, Uzbekistan, Kazakhstan, Iran and Turkey. Mesoamerican archaeology abounds in Mexico, Belize, Guatemala and Honduras for people with an interest in Latin American

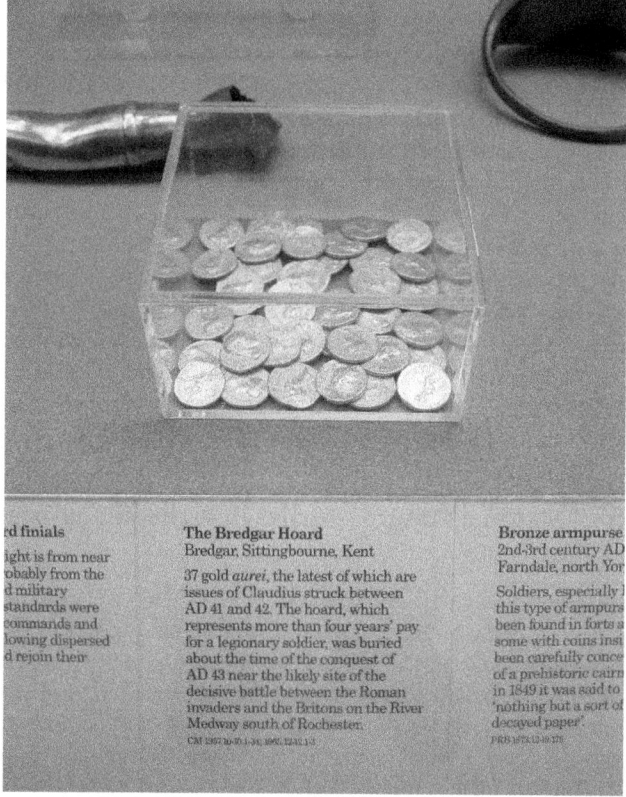

Figure 13.2 Special-interest displays, such as this one of the Bredgar Hoard of Roman coins in the British Museum, appeal to niche markets (Photo: Dallen J. Timothy)

heritage and is a popular attraction both for serious heritage visitors and casual tourists on cruise-based or resort-based day tours.

The most diehard Viking enthusiasts would probably be willing to undertake the arduous journey to Hvalsey settlement (Barr, 2019). The most intrepid biblical aficionados might also be willing to take a risk in visiting ancient Ur, Ninevah and Babylon in present-day Iraq (Guarasci, 2015; Myers et al., 2011; Nawar, 2014). We need more understanding about people's motivations for visiting mainstream archaeological sites, as well as those people who are willing to take a risk to satisfy their historical curiosity. The niche tourism focus that is so well researched in general tourism is lacking in the area of heritage tourism, and even within archaeology tourism there is a need to understand distinct market segments better to be able to offer more satisfying experiences and to understand their impacts. As well, different types of archaeology-based tourism may need different considerations regarding site management, visitor management, impact management and interpretation based on the level of sensitivity of the archaeology and the extent to which a site is attractive for tourists.

Archaeologies and Tourism Types

Several of the chapters in this book have noted the salient role of archaeology in several different types of tourism, yet we know very little about the details. The two main types of tourism discussed in this book are religious tourism and volunteer tourism. Both of these have very clear connections to the work of archaeologists (Kaminski et al., 2011; Koren-Lawrence & Collins-Kreiner, 2019; Neveu, 2010), but what about other types of tourism where the relationship between tourism and archaeology might not be as obvious?

The introductory chapter briefly described the role of artifacts and ancient traditions in agritourism, at least as it manifests in several locations throughout the world. Many vineyard terraces in the Levant, olive terraces and olive oil-associated equipment around the Mediterranean, and the irrigation canals and farming landscapes of the ancient Hohokam society of southern Arizona (USA) are of interest to both archaeologists and tourists. The terraced rice paddies of East and Southeast Asia are of ancient origins but continue to maintain their production function today. They are important tourist attractions, some of which have also been inscribed on the World Heritage List (Guimbatan & Baguilat, 2006; Sun et al., 2011). The domestication of food items in the Mesopotamian cultural hearth thousands of years ago has significant potential as a focus of tourism development, although given the current geopolitical conditions in Iraq and Syria, it will be some time before these important ancient innovations can be utilized for agriculture-based heritage tourism.

Also mentioned briefly at the outset of the book were the associations between archaeology, sport tourism and spa tourism. Both of these forms of tourism are growing niches today, and while many sport and spa enthusiasts are unlikely to be acutely interested in archaeology, there is probably a segment that would have an interest in the history of their pastimes. Thus, the archaeological heritage element of these activities could provide a more holistic tourism product that goes deeper than the demand for archaeological heritage in general (see Figure 13.3).

Solidarity tourism occurs when people travel to support a cause, usually one of a social justice nature. It is an extremely political form of tourism and relies on opposing narratives of 'us versus them'. For example, within the Palestine-Israel conflict, there is a significant solidarity movement on both sides. The pro-Israel faction is composed mostly of certain Jewish groups from the diaspora and particular evangelical Christian groups from North America and Europe. Christian participants visit sacred sites but also undertake activities that demonstrate support for Israel and Zionism, such as attending events, demonstrations and donating money (Belhassen, 2009; Ron & Timothy, 2019). Solidarity tourists for the Palestinian plight visit the occupied Palestinian Territories to show solidarity with these people's predicament and to petition Israeli authorities to ease up on their treatment of Palestinians and return occupied lands. Many of these tourists are political activists from all over the world

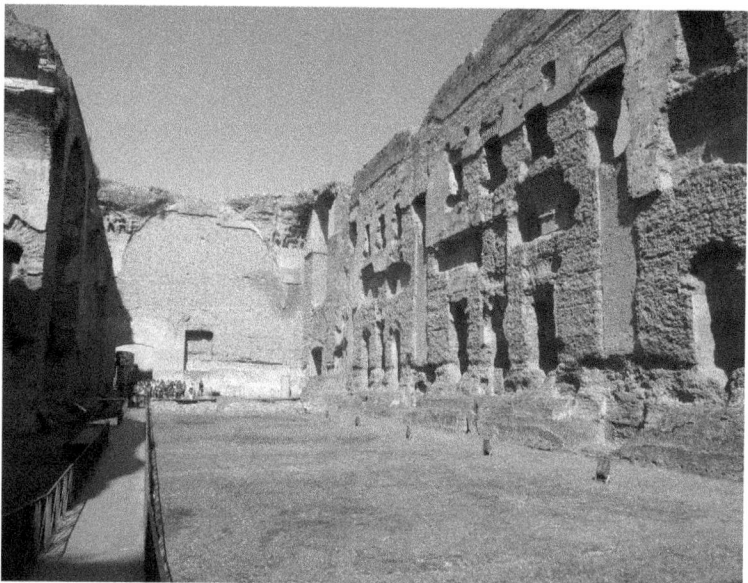

Figure 13.3 Spas were ubiquitous during the Roman Empire, and many, such as Terme di Caracalla in Rome, now serve as important archaeological attractions (Photo: Dallen J. Timothy)

participating in so-called 'justice tours' (Higgins-Desbiolles, 2016; Kassis *et al.*, 2016). There is also a large segment of Christian tourists who desire to show support for Palestinians who are locked behind the 'security wall' (Keating, 2007; Ron & Timothy, 2019). While archaeology has yet to play a significant part in solidarity tourism's efforts specifically, it has considerable potential to be used to justify partisan claims or to tell the story that each side wishes to tell. This can be seen in many parts of the world, including in the Israel-Palestine conflict where museums and archaeological projects in each territory utilize artifacts to establish their own claims of legitimacy. There is much scope to learn more about the political role of archaeology in solidarity tourism contexts everywhere.

Overtourism and Physical Damage

Archaeological remains are a non-renewable resource. Once they are gone, they are gone forever, and they remain one of the most sensitive and delicate tourism assets. While tourism is not the only perpetrator of harm to material culture, it certainly increases the potential for destruction. Countless instances have been recorded of direct and indirect, intentional and unintentional damage to tangible heritage through the actions of tourists and through the growth of tourism (Timothy & Boyd, 2006).

Overall, today's travelers are more aware of their ecological and cultural footprint, and service providers are greener in their approaches to tourism (Reddy & Wilkes, 2015; Séraphin & Nolan, 2019). While international agreements, national legislations, improved education, increased environmental consciousness, and higher levels of outreach by heritage managers have made a difference during the past half century, tourism continues to grow and tourists continue to want to leave their mark. It is incumbent upon archaeologists and cultural resource managers to formulate policies that will prevent physical wear and tear and deliberate damage. However, rules and policies are not enough; enforcement is key, and many places lack the human resources, budgets, or tools to be able to enforce good protective policies. Innovation is needed in the heritage management arena to counteract the growing numbers of visitors who want to experience the archaeological past.

There are ongoing debates and discussions about 'overtourism' in many crowded historic cities, such as Barcelona, Venice, Amsterdam and Prague and how it can best be overcome (Dodds & Butler, 2019). Suggestions are many, but solutions are few. Historic urban centers in Barcelona and Venice suffer excessively through over-visitation, particularly during tourist high season. Residents no longer welcome tourists; crowdedness permeates all aspects of urban life – work, home and leisure. Little is known, however, about the long-term physical impacts of overtourism on the historic environment of cities in Europe and Asia where unbridled tourism growth is a problem in many ancient cities, such as

Kyoto and Beijing, and at many archaeological localities, such as the Great Wall at Badaling (near Beijing), Macau's historic center, and the historic city of Lijiang. Despite rampant over-visitation in European and Asian cities, which in many cases has exceeded plausible carrying capacities, most tourism agents, destination management organizations, and other intermediaries continue to market these locations to mass tourists. Demarketing has been recommended as one way of mitigating some of these problems (Marcotte & Bourdeau, 2012). Demarketing entails discouraging some consumers from visiting, either permanently or only temporarily, and is believed to be an effective tool in managing crowds and tourism impacts in archaeological areas and historic localities (Boyd & Timothy, 2006; Li *et al.*, 2017; Poria, 2013; Soliman, 2010; Timothy & Boyd, 2015).

One common tool for demarketing is virtual reality technology. According to Arnold and Kaminski (2014), some 90% of all tourists visit 10% of the world's heritage attractions. Thus, the most famous sites bear the majority of the burden of overtourism. Arnold and Kaminski believe that technology, including virtual museums and virtual archaeological sites, which can be visited online at home, can help alleviate over-visitation and its impacts on the environment and on the visitor experience. For example, at the Anne Frank House, which is one of the most popular and crowded attractions in an already crowded historic city (Amsterdam), managers have utilized virtual reality as a potential means of demarketing to alleviate some of the problems of overcrowding (Hartmann, 2013; Poria, 2013). Managers have created online tours and encouraged interested parties to tour the museum virtually as an alternative to visiting the location in person (Hartmann, 2013). Perhaps such approaches are warranted at overvisited archaeological sites.

Although we know in general terms the negative implications of too much tourism, because each context is unique with variables that might not exist elsewhere, we need to understand specific impacts on specific types of resources and in different localities. Despite periodic challenges to tourism growth owing to security threats, economic downturns and political instability, there appears to be no end in sight to the growth of global tourism; thus, these concerns are particularly relevant today. It is compulsory for heritage managers and archaeologists to develop innovative means to make tangible heritage more resilient at a time when tourism continues to put increasing pressure on historic environments.

Indigeneity, Archaeology and Descendant Communities

As discussed in other parts of this tome, archaeology is particularly important for descendant communities, especially indigenous people. It has the power to deepen the roots of a people in the place where they live and to provide intergenerational continuity in an increasingly rootless and standardized world. Recent research has acknowledged that tourism has the potential to help empower indigenous people and other descendant societies. By properly planning and managing tourism according to a bottom-up, community-based approach, residents can take control of their socioeconomic futures rather than rely solely on outside control.

When communities take ownership of the problems and benefits associated with tourism, they become psychologically empowered. When tourism brings employment and other economic advantages to the people who want to benefit from it, which is the aim of efforts such as pro-poor tourism and participatory development, destination communities become economically empowered. When native people take pride in their cultural heritage and desire to share it with others through tourism, on their own terms of course, intra-community solidarity grows, and they become socially empowered. When decision-making derives from the grassroots level and as autochthonous societies have the power to welcome tourism or to resist it, they are becoming politically empowered (Scheyvens, 2002).

A critical part of indigenous empowerment relates to cultural property rights. Unfortunately, tourism has not always respected the cultural rights of native peoples. It has in the past frequently appropriated indigenous artifacts and other elements of heritage for profit. Fortunately, during the past half century, conditions have changed in many places, so that native peoples now have a louder voice in how their archaeological record will be used for tourism, including the pillaging and sale of artifacts and the production of replica objects that are meant to represent the tangible culture of natives. In the United States, significant changes have taken place among Native American groups where they are now much more empowered to maintain control over cultural representations in tourism (Nyaupane *et al.*, 2006; Swanson & DeVereaux, 2012). Nonetheless, in the US and many other countries, the use of indigenous culture by non-indigenous tourism promoters remains an extremely delicate topic that needs more evaluation. Increased action research in this area can help empower descendant communities, remove them from the socioeconomic margins of society, and strengthen their group identity.

Future Considerations

In addition to the issues highlighted throughout the volume, there are numerous other matters of concern that need to be addressed through additional research. Two of these, namely climate change and technology, are addressed below.

Climate change

Although most people associate the problems of climate change with biotic systems and living organisms, climate and environmental changes have substantial implications for the built environment, including archaeological remains (Hall, 2016). There is now a vast literature on the effects of climate change on tourism, as well as tourism's contribution to climate change (e.g. Hall *et al.*, 2011; Kaján & Saarinen, 2013; Scott *et al.*, 2012, 2019). Most of the extant research deals with transportation, resource management, infrastructure development, destination growth, activity and behavioral changes, tourist flows, and destination resilience and adaptation. Although there is a growing literature by climate scientists that examines climate change effects on historic and urban environments, environmental change in the realm of archaeology and heritage tourism has not received as much attention as it deserves.

The effects of climate change on the historic built environment, both above ground and underground, are manifold (Hall *et al.*, 2016). The most prominent expected effects are more frequent and increasingly intense storm activity, increased rainfall in some locations, rising sea levels, more instances of extreme events (e.g. floods and heatwaves), coastal erosion, changes in air and soil temperatures and relative humidity, increased soil moisture, augmented invasive species and pests, and intensified solar radiation (Cassar, 2005; Hall, 2019).

Older building materials and unearthed artifacts are particularly vulnerable to many of these changes, especially moisture content and salt crystallization through increased rainfall (Cassar, 2005). Soil composition and moisture content are particularly worrisome to archaeologists, who realize these soil effects of climate change will probably upset 'the equilibrium conditions under which the sites have been preserved for so long' (Cassar, 2005: 5). Cassar (2005: 5–6) also notes that changes in rainfall patterns and temperatures that 'may not be perceived as a major threat to modern buildings, are likely to have dramatic effects on buried or exposed archaeological sites'. Excess moisture is problematic, but so is an absence of moisture. Drying earth can undermine stratigraphy and result in soil cracking and ground collapse.

Flooding, particularly flash floods, is an increasingly problematic result of climate change and affects archaeology in both soil moisture, erosion and washing away artifacts, and the submerging of artifacts in stagnant water (Fernandes, 2016). Increased solar radiation raises challenges to light control in museums and to colored mosaics exposed to outdoor weather, and temperature variations can speed deterioration of certain materials. Excess plant growth and the introduction of invasive species are also a major concern, as vegetative cover and animal pests may directly impact buried sediments and artifacts. As well, deep root penetration is known to fracture delicate ruins and material objects and exacerbate erosion (Cassar, 2005). These outcomes individually or collectively will affect tourism supply and demand, so it is critical for researchers to continue their efforts to promote resilience in the face of environmental challenges.

Archaeology, tourism and technological innovation

Archaeologists have been receptive to technological changes in their scientific toolkits. Many now use DNA tests that were unavailable just two decades ago. Drone technology allows excavation surveyors to take clearer photographs and acquire more detailed depictions of dig sites (Campana, 2017; Hill, 2019). From a resource and visitor management perspective, smart technology has now made it possible to track visitors in archaeological parks and other heritage areas using smartphones or other handheld GPS tracking devices and drone-generated areal images of visitor crowding and spatial behavior (Alexandridis *et al.*, 2019; Garzia *et al.*, 2018; Shoval & Ahas, 2016). This has the potential to help alleviate bottlenecks during busy times of the day, tabulate visitor use in sensitive areas, track the most popular routes and trails, and assist in understanding where best to place interpretive media.

Much research has been done by the tourism industry in collaboration with knowledge enterprises to enhance the visitor experience at archaeological sites in various ways. Mobile phone apps and GPS technology are now widely utilized in archaeological heritage areas to provide information, interactive learning experiences, and entertainment. Geocaching games and other types of gamification are becoming increasingly popular in heritage and archaeological settings through the GPS technology in mobile devices (Etxeberria *et al.*, 2012). Maps, information, guides and even interactive experiences are now available through mobile devices or internet-based websites. Likewise, QR codes have become mainstream links to information sources at many archaeological sites throughout the world. China is particularly astute in providing QR codes in heritage areas, which visitors can easily scan with their mobile devices (Figure 13.4). This provides online information about the site and in fact has the capacity to provide far more information than traditional interpretive signage and placards; it also requires less physical maintenance and fewer changes to the interpretive message.

Virtual reality and augmented reality (AR) are now mainstream concepts in archaeology-based tourism (Han *et al.*, 2018; Njerekai, in press; Rueda-Esteban, 2019; tom Dieck & Han, 2019). Many museums and archaeological sites have embraced the notion of AR (Figure 13.5). At present, this technology has two primary versions of utility. First, AR can use GPS-based mobile applications to replace 2D map-based navigation and effect real-time navigation into the real environment (tom Dieck & Han, 2019). Second, AR is used to overlay imagery onto points of interest in archaeological environments, to 'bring history back to life' by overlaying historical photographs onto historic sites or adding designed imagery to illustrate what missing parts of an archaeological site or monument would have looked like in its original condition (Fusté-Forné, in press; tom Dieck & Han, 2019; tom Dieck & Jung, 2017). For archaeology, AR

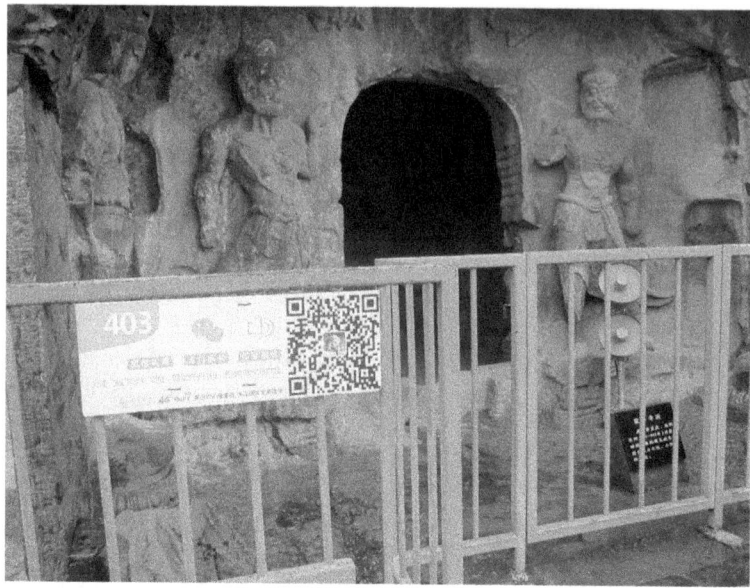

Figure 13.4 Like this setting at Longmen Caves, China, QR Codes and other technologies have become a commonplace interpretive tool in recent years (Photo: Dallen J. Timothy)

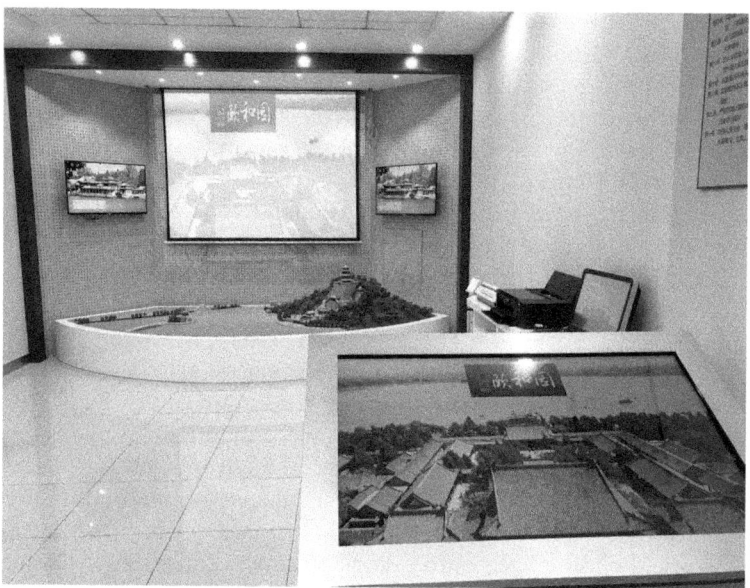

Figure 13.5 Augmented reality helps 'reconstruct' ruins digitally or allows visitors to see how archaeological sites might have looked during different periods of history (Photo: Dallen J. Timothy)

can be implemented to create virtual reconstructions of elements of the built environment that no longer exist based upon research and knowledge about what sites looked like before they became ruins or how they appeared at different times in history (Rueda-Esteban, 2019). This 3D AR perspective is particularly important in archaeological areas where many stratified layers of ruins lie upon one another or where significant portions of built structures are physically missing.

Information communication technology (ICT) is becoming increasingly important in people's lives, so it stands to reason that it will also become essential in how they interact with archaeological heritage. This raises many important questions that beg further research. For example, with growing scientific interest in artificial intelligence (AI), it seems that the future of some aspects of the tourist-archaeology interface may be influenced by AI. Interpretive programs may have to consider AI in how they deliver information, answer questions, and provide 'edutaining' visitor experiences.

Final Word

The aim of this book was to uncover many of the relationships between tourism and archaeology through the interdependent lenses of both fields of study. Tourism scholars and archaeologists both have contributed to this volume and provided unique perspectives on identity and place politics, visitor management, tourism impacts, conservation and protection, economic rationale, privatization, interpretation, public archaeology, volunteering, sustainable marketing and many other issues of critical relevance today.

The description of a few critical issues in this concluding chapter only begins to scratch the surface of what we know and what we do not know. There is much more work to be done. Our hope is that this book has raised many questions and highlighted many points to debate, so that we can continue to deconstruct and understand the multitudinous relationships between tourism and archaeology. This is an important exercise for tourism specialists, archaeologists and other cultural resource managers, and the communities where archaeology-based tourism is most prevalent.

Archaeological remains have long been one of the most outstanding tourism attractions in the world, and they will likely continue to be far into the future. Yet, they remain one of the most literally and metaphorically sensitive assets for tourism. The question now is how archaeologists, tourism specialists, community development experts, service providers and other key stakeholders will shape resources and destinations to be more resilient to change and use by tourism. Likewise, how will they govern the tourism industry to be innovative, adaptive, and resilient in light of the need for more responsible corporate, destination and tourist behavior (Lew, 2014)? Our hope is that the content of this book provides concepts, ideas, and experiences that can help develop sustainable tourism that both respects archaeological resources and uses them wisely.

At the close of this book, it seems pertinent to mention the fire destruction of the roof and spire of Notre Dame Cathedral in Paris and its associated archaeology, which occurred on 15 April 2019. Tourists and Parisians alike watched in horror as a symbolic monument was ravaged by fire and wondered whether the 800-year-old cathedral would survive. This was truly a tragedy, not just for Paris, but also for the entire world. In response to this disaster, individuals and organizations pledged nearly 1 billion euros to rebuild the iconic shrine. The destruction was not deliberate, but it caused archaeologists to think about why a monument would receive such huge sums of money for its restoration. Is it because the cathedral is branded a UNESCO World Heritage Site, or is it because of its quintessential place in the heritagescape of the capital city of the most visited country in the world? When monuments and archaeological sites in other parts of the world with equivalent beauty are destroyed, deliberately or not, they rarely receive the same level of mournful attention, nor do they receive similar quantities of cash to be saved and cherished for future generations. What is it about Notre Dame Cathedral that compelled some of Europe's wealthiest people to donate towards its restoration? Do we value a certain archaeological heritage at the exclusion of others? The destruction of archaeological sites in Nineveh and Nimrud, Iraq and Palmyra, Syria, by ISIL in 2014–2015 was perhaps even more devastating given its intentionality and its permanence, yet these sites did not garner as much 'sponsored interest' as the cathedral did in Paris. It is critical to consider archaeological remains as priceless heritage that is precious to all of humanity, valuable assets for cultural tourism and the foundation of national pride. Tourism has a role to play in financing, protecting and restoring the archaeological record. More research is needed to understand how this can be done more effectively throughout the world.

References

Alexandridis, G., Chrysanthi, A., Tsekouras, G.E. and Caridakis, G. (2019) Personalized and content adaptive cultural heritage path recommendation: An application to the Gournia and Çatalhöyük archaeological sites. *User Modeling and User-Adapted Interaction* 29 (1), 201–238.

Arnold, D. and Kaminski, J. (2014) Cultural heritage tourism and the digital future. In J. Kaminski, A.M. Benson and D. Arnold (eds) *Contemporary Issues in Cultural Heritage Tourism* (pp. 261–282). London: Routledge.

Barr, S. (2019) Cultural heritage, or how bad news can also be good. In N. Sellheim, Y.V. Zaika and I. Kelman (eds) *Arctic Triumph: Northern Innovation and Persistence* (pp. 43–57). Cham, Switzerland: Springer.

Belhassen, Y. (2009) Fundamentalist Christian pilgrimages as a political and cultural force. *Journal of Heritage Tourism* 4 (2), 131–144.

Berry, S. and Shephard, G. (2001) Cultural heritage sites and their visitors: Too many for too few? In G. Richards (ed.) *Cultural Attractions and European Tourism* (pp. 159–171). Wallingford: CABI.

Boyd, S.W. and Timothy, D.J. (2006) Marketing issues and World Heritage Sites. In A. Leask and A. Fyall (eds) *Managing World Heritage Sites* (pp. 53–66). Oxford: Butterworth Heinemann.

Cameron, C.M. and Gatewood, J.B. (2008) Beyond sun, sand and sea: The emergent tourism programme in the Turks and Caicos Islands. *Journal of Heritage Tourism* 3 (1), 55–73.

Campana, S. (2017) Drones in archaeology: State-of-the-art and future perspectives. *Archaeological Prospection* 24 (4), 275–296.

Cassar, M. (2005) *Climate Change and the Historic Environment.* London: University College London.

Catalani, A. and Ackroyd, T. (2013) Inheriting slavery: Making sense of a difficult heritage. *Journal of Heritage Tourism* 8 (4), 337–346.

Deacon, J. (2006) Rock art conservation and tourism. *Journal of Archaeological Method and Theory* 13 (4), 376–396.

Dodds, R. and Butler, R. (eds) (2019) *Overtourism: Issues, Realities and Solutions.* Berlin: De Gruyter.

Duval, M. and Smith, B. (2013) Rock art tourism in the uKhahlamba/Drakensberg World Heritage Site: Obstacles to the development of sustainable tourism. *Journal of Sustainable Tourism* 21 (1), 134–153.

Etxeberria, A.I., Asensio, M., Vicent, N. and Cuenca, J.M. (2012) Mobile devices: A tool for tourism and learning at archaeological sites. *International Journal of Web Based Communities* 8 (1), 57–72.

Fernandes, F. (2016) Built heritage and flash floods: Hiking trails and tourism on Madeira Island. *Journal of Heritage Tourism* 11 (1), 88–95.

Fusté-Forné, F. (in press) Mapping heritage digitally for tourism: An example of Vall de Boí, Catalonia, Spain. *Journal of Heritage Tourism.*

Garzia, F., Lombardi, M. and Papi, L. (2018) Analysis and data acquisition methodology based on flying drones for the implementation of the internet of everything to smart archaeological areas. *International Journal of Heritage Architecture* 2 (3), 383–394.

Guarasci, B.L. (2015) The national park: Reviving Eden in Iraq's marshes. *The Arab Studies Journal* 23 (1), 128–153.

Guimbatan, R. and Baguilat Jr, T. (2006) Misunderstanding the notion of conservation in the Philippine rice terraces–cultural landscapes. *International Social Science Journal* 58 (187), 59–67.

Hall, C.M. (2016) Heritage, heritage tourism and climate change. *Journal of Heritage Tourism* 11 (1), 1–9.

Hall, C.M. (2019) Tourism and climate change in the Middle East. In D.J. Timothy (ed.) *Routledge Handbook on Tourism in the Middle East and North Africa* (pp. 199–209). London: Routledge.

Hall, C.M., Baird, T., James, M. and Ram, Y. (2016) Climate change and cultural heritage: Conservation and heritage tourism in the Anthropocene. *Journal of Heritage Tourism* 11 (1), 10–24.

Hall, C.M., Scott, D. and Gössling, S. (2011) Forests, climate change and tourism. *Journal of Heritage Tourism* 6 (4), 353–363.

Han, D.-I., tom Dieck, M.C. and Jung, T. (2018) User experience model for augmented reality applications in urban heritage tourism. *Journal of Heritage Tourism* 13 (1), 46–61.

Hartmann, R. (2013) The Anne Frank House in Amsterdam: A museum and literary landscape goes virtual reality. *Journalism and Mass Communication* 3 (10), 625–644.

Higgins-Desbiolles, F. (2016) Wall off from the world: Palestine, tourism and resisting occupation. In R. Isaac, C.M. Hall and F. Higgins-Desbiolles (eds) *The Politics and Power of Tourism in Palestine* (pp. 178–194). London: Routledge.

Hill, A.C. (2019) Economical drone mapping for archaeology: Comparisons of efficiency and accuracy. *Journal of Archaeological Science: Reports* 24, 80–91.

Jolliffe, L. (ed.) (2013) *Sugar Heritage and Tourism in Transition*. Bristol: Channel View Publications.

Jordan, L.-A. and Jolliffe, L. (2013) Heritage tourism in the Caribbean: Current themes and challenges. *Journal of Heritage Tourism* 8 (1), 1–8.

Kaján, E. and Saarinen, J. (2013) Tourism, climate change and adaptation: A review. *Current Issues in Tourism* 16 (2), 167–195.

Kaminski, J., Arnold, D.B. and Benson, A.M. (2011) Volunteer archaeological tourism: An overview. In A.M. Benson (ed.) *Volunteer Tourism: Theoretical Frameworks and Practical Applications* (pp. 157–174). London: Routledge.

Kassis, R., Solomon, R. and Higgins-Desbiolles, F. (2016) Solidarity tourism in Palestine: The alternative tourism group of Palestine as a catalyzing instrument of resistance. In R. Isaac, C.M. Hall and F. Higgins-Desbiolles (eds) *The Politics and Power of Tourism in Palestine* (pp. 37–53). London: Routledge.

Keating, M. (2007) Plan to take a life-changing trip to Palestine in 2007. *The Washington Report on Middle East Affairs* 26 (2), 69–70.

Koren-Lawrence, N. and Collins-Kreiner, N. (2019) Visitors with their 'backs to the archaeology': Religious tourism and archaeology. *Journal of Heritage Tourism* 14 (2), 138–149.

Lew, A.A. (2014) Scale, change and resilience in community tourism planning. *Tourism Geographies* 16 (1), 14–22.

Li, S.C., Robinson, P. and Oriade, A. (2017) Destination marketing: The use of technology since the millennium. *Journal of Destination Marketing & Management* 6 (2), 95–102.

Logan, W. (2013) Australia, indigenous peoples and World Heritage from Kakadu to Cape York: state party behaviour under the World Heritage Convention. *Journal of Social Archaeology* 13 (2), 153–176.

Marcotte, P. and Bourdeau, L. (2012) Is the World Heritage label used as a promotional argument for sustainable tourism? *Journal of Cultural Heritage Management and Sustainable Development* 2 (1), 80–91.

Meskell, L. (2013) UNESCO's World Heritage Convention at 40: Challenging the economic and political order of international heritage conservation. *Current Anthropology* 54 (4), 483–494.

Meskell, L., Liuzza, C., Bertacchini, E. and Saccone, D. (2015) Multilateralism and UNESCO World Heritage: Decision-making, States Parties and political processes. *International Journal of Heritage Studies* 21 (5), 423–440.

Myers, S.L., Farrell, S. and Fukada, S. (2011) A tour of Iraq's ancient sites. *The New York Times*, 2 January, 2011. See https://atwar.blogs.nytimes.com/2011/01/02/a-tour-of-iraqs-ancient-sites/ (accessed 22 March 2019).

Nawar, A.S. (2014) Insights into the main difficulties of achieving sustainable development of tourism in Iraq. *Analele Universitanii din Oradea-Seria Geografie* 1, 32–43.

Neveu, N. (2010) Islamic tourism as an ideological construction: A Jordan study case. *Journal of Tourism and Cultural Change* 8 (4), 327–337.

Njerekai, C. (in press) An application of the virtual reality 360° concept: The Great Zimbabwe monument. *Journal of Heritage Tourism*.

Nyaupane, G.P., White, D.D. and Budruk, M. (2006) Motive-based tourist market segmentation: An application to Native American cultural heritage sites in Arizona, USA. *Journal of Heritage Tourism* 1 (2), 81–99.

Poria, Y. (2013) The four musts: See, learn, feel, and evolve. *Journal of Heritage Tourism* 8 (4), 347–351.

Reddy, M.V. and Wilkes, K. (eds) (2015) *Tourism in the Green Economy*. London: Routledge.

Ribaudo, G. and Figini, P. (2017) The puzzle of tourism demand at destinations hosting UNESCO World Heritage Sites: An analysis of tourism flows for Italy. *Journal of Travel Research* 56 (4), 521–542.

Ron, A.S. and Timothy, D.J. (2019) *Contemporary Christian Travel: Pilgrimage, Practice and Place*. Bristol: Channel View Publications.

Rueda-Esteban, N.R. (2019) Technology as a tool to rebuild heritage sites: The second life of the Abbey of Cluny. *Journal of Heritage Tourism* 14 (2), 101–116.

Scheyvens, R. (2002) *Tourism for Development: Empowering Communities*. Harlow: Prentice Hall.

Scott, D., Gössling, S. and Hall, C.M. (2012) *Climate Change and Tourism: Impacts, Adaptation and Mitigation*. London: Routledge.

Scott, D., Hall, C.M. and Gössling, S. (2019) Global tourism vulnerability to climate change. *Annals of Tourism Research* 77, 49–61.

Séraphin, H. and Nolan, E. (eds) (2019) *Green Events and Green Tourism: An International Guide to Good Practice*. London: Routledge.

Schmitt, T.M. (2009) Global cultural governance. decision-making concerning World Heritage between politics and science. *Erdkunde* 63 (2), 103–121.

Scholze, M. (2008) Arrested heritage: the politics of inscription into the UNESCO World Heritage List: The case of Agadez in Niger. *Journal of Material Culture* 13 (2), 215–231.

Shoval, N. and Ahas, R. (2016) The use of tracking technologies in tourism research: the first decade. *Tourism Geographies* 18 (5), 587–606.

Soliman, D.M. (2010) Managing visitors via demarketing in the Egyptian World Heritage Site: Giza Pyramids. *Journal of the Association of Arab Universities for Tourism and Hospitality* 7 (1), 15–20.

Sun, Y., Min, Q., Shi, J. and Jiang, Y. (2011) Terraced landscapes as a cultural and natural heritage resource. *Tourism Geographies* 13 (2), 328–331.

Swanson, K.K. and DeVereaux, C. (2012) Culturally sustainable entrepreneurship: A case study for Hopi tourism. In K.F. Hyde, C. Ryan and A.G. Woodside (eds) *Field Guide to Case Study Research in Tourism, Hospitality and Leisure* (pp. 479–494). Bingley: Emerald.

Timothy, D.J. (2014a) Contemporary cultural heritage and tourism: Development issues and emerging trends. *Public Archaeology* 13 (3), 30–47.

Timothy, D.J. (2014b) Views of the vernacular: tourism and heritage of the ordinary. In J. Kaminski, A. Benson, and D. Arnold (eds) *Contemporary Issues in Cultural Heritage Tourism* (pp. 32–44). London: Routledge.

Timothy, D.J. (2018) Making sense of heritage tourism: Research trends in a maturing field of study. *Tourism Management Perspectives* 25, 177–180.

Timothy, D.J. and Boyd, S.W. (2006) Heritage tourism in the 21st century: Valued traditions and new perspectives. *Journal of Heritage Tourism* 1 (1), 1–17.

Timothy, D.J. and Boyd, S.W. (2015) *Tourism and Trails: Cultural, Ecological and Management Issues*. Bristol: Channel View Publications.

tom Dieck, M.C. and Han, D.-I. (2019) Tourism and augmented reality: Trends, implications and future directions. In D.J. Timothy (ed.) *Handbook of Globalization and Tourism* (pp. 235–246). Cheltenham: Edward Elgar.

tom Dieck, M.C. and Jung, T.H. (2017) Value of augmented reality at cultural heritage sites: A stakeholder approach. *Journal of Destination Marketing & Management* 6 (2), 110–117.

Wurz, S. and van der Merwe, J.H. (2005) Gauging site sensitivity for sustainable archaeotourism in the Western Cape Province of South Africa. *South African Archaeological Bulletin* 60 (181), 10–19.

Index

3D technology 73, 156, 175–176, 177, 178, 180–181, 219
  See also virtual reality; augmented reality
360° technology 179–180

access 26, 31, 35, 71, 72, 156, 162, 163, 193, 206
Acropolis (Greece) 106, 168
Afqa Cave (Lebanon) 109–110
Afghanistan 14, 125
Africa 16, 92, 98, 138, 143, 146, 189
agriculture 10, 12, 14, 19, 89, 121, 157, 208, 211
agritourism 10, 211
Al-Qaeda 125–126
Altamira (Spain) 161
alternative tourism 11, 88–89
American Anthropological Association 161
Angkor Wat (Cambodia) 6, 12, 31, 62, 140, 155
antiquities trade 43, 135–136, 142, 144, 146–147
  See also illicit trade in artifacts
Antonine Wall (Scotland) 182
archaeological parks 5, 6, 8, 12, 27–28, 35, 76–79, 128, 139, 177
archaeological resource management (ARM) *See* Cultural Resource Management
archaeological tourism *see* archaeotourism
archaeotourism 5, 6, 8, 10, 27, 30, 35–36, 69, 71–73, 75, 82, 93, 121, 152, 153–154, 161, 177, 189, 190, 193, 196, 205, 207, 211
artificial intelligence (AI) 219
Asia 1, 10, 98, 136–137, 142, 189, 211, 213
Assyrian civilization 3, 92
Aubechies (Belgium) 28, 35

augmented reality 73, 83, 179–180, 217–219
  See also virtual reality
Australia 32, 72, 98, 170, 176
Austria 63, 208
authenticity 20, 29, 65, 69, 70–72, 73, 4, 83, 117, 131, 134, 141–142, 145–146, 147, 154, 160–161, 171–172, 177, 179

Baalbek (Lebanon) 108, 111, 210
Baghdad Archaeological Museum 153
Bamiyan Valley (Afghanistan) 14, 125
Bangladesh 74
Bath (England) 10
Belgium 11, 28, 35, 73
Belize 96, 138, 145, 210
Bible 96–97, 115, 142, 210–211
Biblical archaeology 3, 96–97, 115–118
Bibracte (France) 28, 35
black market 14, 126, 136, 139, 143
  See also illicit trade in artifacts
Bosnia and Herzegovina 108
Bosra (Syria) 28
branding, heritage 4, 5, 9–10, 18, 44, 69, 72, 73–75, 82, 83, 115, 190, 191–192, 205–207
  See also marketing
British Museum 167, 210
Bronze Age 45, 97, 99
budgets *see* funding
Buddha 117, 125
Buddhism 75
built environment 3, 12, 19, 20, 28, 130, 162, 216, 219
  See also cultural landscapes
Burra Charter 42, 172
Byzantine period 110, 113, 116

Cambodia 6, 12, 31, 54, 60, 61, 138, 139, 140, 142, 155, 191, 207
Canaanites 111, 113

Canada 27, 74, 173–174, 181, 210
Canadian Tourism Commission 74
Caribbean 69, 76, 209
carrying capacity 15, 131, 155, 193, 214
Çatalhöyük (Turkey) 6, 57, 113
  See also Turkey
cataloging 92
cemeteries 45, 106, 209
Charter for the Interpretation and Presentation of Cultural Heritage Sites (ICOMOS) 171–172
Charter for the Protection and Management of the Archaeological Heritage (ICOMOS) 171
Charter on the Built Vernacular Heritage (ICOMOS) 171
Chichen Itza (Mexico) 11, 61–62, 76–78, 153, 207
China 5, 6, 10, 14, 137, 138, 139, 154, 159, 162, 206–207, 217
Christianity/Christians 1, 16, 46, 96–97, 107, 109–111, 113, 116, 118, 126, 140, 210, 212–213
Church of the Holy Sepulchre 10, 111
churches 106, 108, 110–111, 112, 114, 116, 118, 208
climate change 169, 205, 215–216
collaboration 36, 71, 72–73, 76, 97, 100, 156, 182, 193
collections/collectors 8, 14, 27, 134–147, 152
colonialism 4, 16, 33, 62, 110, 136, 138, 171
commercialization *see* commoditization
commoditization 29, 62, 152, 178, 192
community archaeology *see* public archaeology
community-based tourism 75, 156–158, 160, 175, 194, 215
conservation 4, 8, 11–12, 26, 27, 28, 31, 34–35, 36, 41, 42, 44–45, 58, 61, 63, 71, 73, 80, 91, 121, 123, 127–128, 131, 138, 152–164, 207
Copán (Honduras) 153
crowding *see* overtourism
cruises 77, 211
cultural landscapes 3, 73, 74, 130, 170, 205, 208
cultural heritage management (CHM) *see* cultural resource management

cultural resource management (CRM) 3, 4, 8, 17, 93, 213
cultural tourism 5, 27, 69, 70, 72, 79, 117, 121, 123, 129, 131, 153, 154, 163, 189, 193, 220
curating/curators 3, 62, 91, 137, 176
Cusco (Peru) 115
Cyprus 74, 97, 210

decolonization 4, 30
Delphi (Greece) 108, 114
demand *see* market, heritage
demarketing 214
Democratic Republic of the Congo 140
democratization of archaeology/heritage 30, 96, 196
Denmark 139
descendant communities 4, 16, 36, 58, 78, 160–161, 208, 214, 215
  See also Indigenous people
destination management/marketing organization (DMO) 80, 117, 214
diasporas 4, 7, 212
digital media 18, 73, 81, 172, 177, 179–183
digs *see* excavations
disasters, natural 93, 121, 183, 189
disempowerment 15, 16, 21, 89, 102
Disneyfication 33, 161, 177
dissonance 28, 117, 118, 175
domestic tourists 27, 70, 74, 79, 80, 98, 144, 162–163, 207
drones 2, 217

Easter Island 145
economic development 8, 28–29, 33, 44, 71–72, 77, 123, 152, 155, 188
  See also sustainable development
ecotourism 87
Ecuador 100
educational programs *see* interpretation
education, formal 10, 93, 95, 98–99, 169
education, informal *see* interpretation
Egypt 1, 6, 12, 31, 97, 113, 122, 136, 138–139, 140, 142, 152, 153, 207, 210
Egyptian pyramids 1, 12, 106
El Salvador 138
employment 26, 42, 47, 65, 74, 157–158, 175
empowerment 15–16, 21, 97, 157–158, 169, 175, 181, 193, 208, 215
enforcement, law 13, 136, 138–139, 141, 145, 147, 162, 207, 213

England 6, 10, 74, 156, 167–168, 170
    See also United Kingdom
English Heritage 91, 176, 189
entertainment 15, 153, 167, 177, 217
entrance fees 8, 44, 46, 54, 57, 63, 153, 159
Estonia 139
ethics 44, 159–161, 175, 178, 196
Europe 1, 3, 10, 14, 16, 45, 55, 113, 136, 138, 139, 187, 189, 192, 212, 213
excavations 3, 6, 8, 10, 28, 29, 31, 33, 34–35, 46, 62, 65, 71, 72, 74, 77, 91–93, 96, 100, 110–111, 115–116, 121, 129, 137, 139, 159, 168, 170, 182, 205, 207

farming see agriculture
field school see service learning
food 10
France 28, 31, 33, 35, 108, 113, 122, 161
funding 8, 32, 42, 46–48, 55, 59, 63, 80, 91, 98, 129, 161, 171, 187, 207

galleries 8, 74
games, as interpretive tool see gamification
gamification 175, 180–182, 217
gender 170–171
geocaching 217
Germany 210
Global Heritage Fund 44
globalization 55, 188, 190
government/state, role 32, 42–45, 55–60, 91, 98, 129, 158, 171, 187, 195
graffiti see vandalism
Grand Tour 1–2, 167
grave robbing see looting
graves see tombs
Great Britain see United Kingdom
Great Wall of China 5, 159, 214
Greece 6, 54, 60, 63–64, 74, 106, 122, 136, 145, 153, 168, 207, 210
Greek Empire 1, 110–111
Greenland 210
Guatemala 75, 138, 210
guidebooks 115, 125, 159, 176
guides 17, 80, 96, 142, 156, 163, 169, 194

Hadrian's Wall (England) 210
Hellenistic Empire see Greek Empire
Herculaneum (Italy) 178–179

Heritage Council (Ireland) 91
heritage, definition 4–5
Heritage New Zealand 91
heritagization 5, 31
Hinduism 1, 107, 113, 118, 123–124
Historic Environment Scotland 79–83
hobby, archaeology as 7, 90, 94–95, 97, 168, 205, 209–210, 212
Holy Land 13, 96–97, 115–117, 136, 140–144
Honduras 75, 153, 210
human rights see social justice

Iceland 139
International Council on Monuments and Sites (ICOMOS) 30, 59, 74, 168, 171–172, 180, 183, 190
identity, national/cultural 14, 30, 36, 71, 74, 77, 107, 115, 123, 126, 172, 186, 208, 215
ideology see politics, heritage/archaeology
illicit trade in artifacts 14, 19, 62, 134–147
    See also looting
impacts of tourism
    destructive 8, 11–12, 19, 26, 28, 29, 31, 45, 72, 80, 117, 121–131, 152, 154–156, 159–160, 162, 180
    economic 11–12, 26, 33, 41–50, 58–59, 71, 74, 117, 154, 157, 160, 195
inauthentic see authenticity
India 1, 4, 6, 13, 63, 75, 92, 113, 118, 123–125, 178, 207
Indiana Jones 167
Indigenous people 4, 16, 32, 35, 76, 138, 157–160, 163, 170–171, 173–175, 205, 214–215
Indonesia 10
industrial archaeology 6
information technology 35
infrastructure development 14, 28, 33, 44, 72, 121, 158, 216
intangible heritage 3–4, 12, 41, 106–107, 113, 114, 157, 159, 169, 174, 179, 180–181, 190, 196
International Council of Museums 137
International Monetary Fund 55
interpretation/interpretive programs 3, 4, 8, 11–12, 15, 17, 27, 29, 32, 34–35, 44, 48, 59, 73, 91, 92–93, 96, 117,

130, 160, 167–182, 186, 193–194, 207, 211
interpretive centers 6, 27, 70, 90, 101, 169, 176, 205
interpretive media/methods 35, 46–47, 71, 78, 175–182, 217–218
inventorying 33
Iran 210
Iraq 14, 92, 125–126, 140, 153, 210–211
Iron Age 46, 99
ISIS (Islamic State in Iraq and Syria)/Daesh 14, 124–126, 140
Islam 16, 107, 109–111, 116, 118, 124, 125–126
Israel 97, 142, 144, 191, 210, 212
Italy 6, 31, 54, 60, 61, 62, 74, 158, 168, 178, 206, 207, 210

Japan 113
Jerusalem 1, 10, 111, 114, 142–144
Jesus Christ 97, 111, 115, 117, 142, 210
Jews *see* Judaism
job creation *see* employment
Jordan 5, 12, 28, 30, 75, 97, 122, 142, 145, 154, 163, 210
Jordan River 113
Jordanian Department of Antiquities 155
Jorvik Viking Centre (England) 177, 210
Judaism 97, 107, 212

Karnak temples (Egypt) 108
Kazakhstan 210
Kilmartin Glen (Scotland) 41, 45–50

L'Anse aux Meadows (Canada) 210
Lascaux (France) 31, 122, 161
Latin America 14, 98, 110, 137, 138, 142, 160, 210
See also Mesoamerica
laws *see* legislation
leakage, economic 48, 192
Lebanon 16, 107–109, 111–118, 127–130, 210
legislation, protective 14, 138–140, 142
leisure 8, 11, 14, 18, 28, 30, 35, 88, 94, 134–135, 139, 163, 205, 209, 212
Levant 96, 136, 211
Liechtenstein 7, 208

living culture *see* intangible heritage
local communities *see* resident populations
looting 12, 14, 28, 43, 62, 123, 126, 135–147, 215
See also illicit trade in artifacts
Luxor (Egypt) 12, 114

Macau 214
Machu Picchu (Peru) 5, 31, 64, 115, 180, 207
Mali 125
managing archaeological resources 14–15, 35, 45, 61, 71, 87–91, 117, 155, 159, 162, 193, 211
See also visitor management
maritime archaeology 72
market characteristics 7, 8–9, 44, 59–60, 70–71, 82–83, 87, 90–91, 92–97, 117, 153–154, 163, 211
market forces 42, 54, 56, 63, 69
marketing 4, 8–9, 34, 64, 69, 71–83, 91, 159, 163, 192, 205, 214
See also branding
mass tourism 8, 11–12, 17, 19, 62, 87–89, 101, 121–123, 193, 209
Maya civilization 10–11, 75, 96, 153, 157
Mecca (Saudi Arabia) 114
media 54, 60, 63, 81, 145, 153, 176, 193
Medieval period 46, 92, 99, 110, 118, 135, 156
Mesoamerica 10, 76–79, 153, 210
Mesopotamia 111, 140, 211
Mexico 11, 54, 60, 61–62, 69, 75–79, 110, 122, 138, 153, 157, 207, 210
microstates 7, 208
Middle Ages 1, 2
Middle East 3, 10, 14, 58, 98, 107, 112, 125–126, 142–143, 153
mobile apps 217
See also smartphones
Moldova 139
Monaco 7
monuments 6, 7, 13, 45, 47, 63, 64, 73, 123, 125, 126, 161, 170, 173, 177, 180, 205, 217, 220
mosques 106, 114, 123–124, 143
motivations for visiting 7, 47, 70, 74, 87–90, 94, 107–108, 162–163, 175, 177, 205, 211

museums 6, 8, 12, 16, 27, 31, 44, 45–50, 54, 57, 62–64, 70, 74, 76–79, 88, 90–91, 94, 100, 110, 113, 125, 129, 135–137, 140–144, 156, 159, 168–169, 176, 178, 193, 205, 209–210, 214, 216–217
Muslims *see* Islam

Nara Document on Authenticity (ICMOS) 171–172
nationalism 14, 15, 20, 29, 30, 74, 123, 167, 186–194
 See also politics, heritage/archaeology
national parks 6, 169
National Historic Landmarks program (USA) 9
National Museum of Scotland 81
National Register of Historic Places (USA) 9
National Trust for Historic Preservation (USA) 91
National Trust for Scotland 81
Native Americans 16, 160, 208, 210, 211
 See also Indigenous people
native peoples *see* Indigenous people
Near Eastern Archaeology 3
neocolonialism 12, 89
neoliberalism 55, 60–61
 See also globalization
Neolithic archaeology/sites 3, 45, 57, 113, 162
Nepal 145
New Age tourism 108
New Zealand 169, 173
Netherlands 62–63, 172, 176–177
Nigeria 145
Nimrud (Iraq) 125
Nineveh (Iraq) 125–126
nomads 208
non-profit organizations (NGOs) 8, 55, 63, 88, 98, 156
Norway 139

Ottoman Empire 110–111, 138, 143
overtourism 162, 189, 205, 213–214

Pacific Islands 208
Palestine 6, 97, 142, 144, 153, 191, 210, 212
Palmyra (Syria) 125–126
Parque Eco-Arqueologico (Mexico) 75–79

participatory archaeology *see* public archaeology
partnerships *see* collaboration
personal heritage 4, 7
Peru 5, 31, 64, 145, 207
Petra (Jordan) 5, 12, 28, 30, 115, 122, 154, 162, 163, 180
Philippines 10
Phoenicians 111, 113
Pilgrimage/pilgrims *see* religious tourism
pillaging *see* looting; *see also* vandalism
planning, tourism 12, 155, 157, 193
political instability 140, 211, 212, 214
politics, heritage/archaeology 9, 15–17, 20, 30–31, 34, 42, 44, 55–56, 74–75, 107, 117, 123–125, 167, 170, 186–194, 205–206, 213
pollution 12–13, 155
Pompeii (Italy) 31, 99, 167–168, 178–179, 180
Portugal 108
poverty reduction *see* pro-poor tourism
power *see* politics
Preah Vihear temple (Cambodia) 191
pre-Columbian cultures 77–79, 110
pre-Hispanic era *see* pre-Columbian cultures
pricing 15
privatization 54–65
pro-poor tourism 8, 172, 192, 215
protection *see* conservation
public archaeology 3, 8, 9, 14, 27, 29, 32, 34–36, 42, 48, 57–58, 62, 63, 87, 97–98, 160–161, 171, 182
public memory *see* values, identity

QR codes 217–218

recreation *see* leisure
religious archaeology 106–118
religious fanaticism 12, 14, 123–126
religious tourism 1, 10, 13, 106–118, 211
remote sensing 2
repatriation of archaeological relics 137, 160
replicas 74, 145–146, 156, 161, 172, 176, 215
reproductions *see* replicas
resilience 208, 214, 216, 219
rescue archaeology 3, 30

resident populations 28, 31, 44, 47–48, 58, 71, 72, 117, 152, 156–157, 160, 171, 172, 194
  See also descendant communities
restoration *see* conservation
revenue *see* funding
ritual landscapes 46
rituals 106, 112–113, 118, 159, 169
rock art 162–163, 173, 174, 208–209
Roman Empire 1, 10–11, 95, 109, 110–111, 113, 136, 139, 172, 182, 210, 212
Romania 139
Romanticism 168
routes *see* trails
Russia 139
Rwanda 175

sacred sites 1, 107–117
scale 6, 26, 46, 49, 57, 98, 205, 207–209
Scotland 10, 41, 45–50, 69, 72, 79–83, 182
  See also United Kingdom
selfies 7
sense of place 4, 69, 83, 97, 158, 209
Serbia 139
serious leisure 94, 209, 211
service learning 93–97
seven wonders of the ancient world 1
shrines 10, 106, 108–110, 112, 114, 118, 220
Silk Road 210
slavery 16, 96, 209
smartphones 179, 217
smuggling *see* illicit trade in artifacts
social justice 88, 89, 160–161, 163, 212
social media 57, 81–83, 159, 196
Society for American Archaeology 161, 177
soil sampling 2, 92
soil sifting 3, 92
solidarity tourism 212–213
Somalia 140
South Africa 162–163
souvenirs 122, 134, 142, 147, 176
spas 10–11, 212
Spain 6, 94–95, 110, 158, 161, 207
sport tourism 212
stakeholders 26, 31, 36, 41, 44, 47, 72, 75, 97, 131, 156, 158, 160, 173, 175
  See also collaboration

stone circles 45–46
Stonehenge (England) 6, 74, 106, 115, 174, 177
sun, sea and sand tourism (SSS) 69, 72, 76, 153, 209
surveys 2, 45, 46, 73, 92, 99, 217
sustainable development 26, 71, 131, 155–156, 192–193, 194
  See also sustainable tourism
sustainable tourism 35, 71, 72, 116, 122, 155, 157, 192
Sweden 139
Switzerland 208
synagogues 118
Syria 14, 28, 125–126, 210

Taj Mahal (India) 5, 13, 115, 125
Taliban 14, 125–126
Temple of Antoninus and Faustina (Italy) 108
temples 1, 12, 63, 106, 108–109, 110–113, 114, 118, 123, 136, 140, 191, 210
Teotihuacan (Mexico) 110, 207
terrorism 14, 19, 121, 125–126, 140
Thailand 6, 191
Theban Necropolis (Egypt) 145
Thomas Cook 2
Tilden's principles of interpretation 169–170
Timbuktu (Mali) 125
tombs 46, 73, 111, 137–138
tour operators 115, 142, 214
trails 7, 18, 71, 72, 80, 89, 92, 112, 128, 169, 217
travel agencies *see* tour operators
Tulum (Mexico) 76–78
Turkey 1, 6, 57, 64, 97, 113, 136–137, 142, 143, 146, 156, 207, 210
Turkmenistan 210
Tutankhamun 161, 167

Ukraine 139
UNESCO 9, 30, 69, 74, 80, 106, 115, 117, 126, 130, 139, 146, 157, 182, 190–191, 195, 206–207
  See also World Heritage Sites
United Kingdom 2, 3, 6, 27, 42, 55, 63, 74, 80, 95–96, 98, 106, 176, 177, 189–190, 207
  See also England, Scotland

United States of America 3, 9, 16, 27, 32, 33, 55, 93, 95, 98, 146, 156, 160, 168, 191, 207, 211, 215
US National Park Service 9, 91, 156, 168
Uzbekistan 210

values, heritage
  aesthetic 42, 71, 129
  cultural 41, 71, 73, 138, 173, 188
  economic 8, 33, 41–50, 59, 107, 154, 160, 188
  historical 42, 124
  identity 42, 77, 97, 123, 129
  political 33, 107, 213
  scientific 42, 48, 138, 171, 180
  social 33, 163
  spiritual 42, 71, 73, 173
Valley of the Kings (Egypt) 31
vandalism 12–13, 122–123, 154–155
Venice Charter (ICOMOS) 168, 171
vernacular heritage 207
Vietnam 139
Viking period 139, 209, 210
virtual archaeology *see* virtual reality
virtual reality 83, 161, 175, 179–180, 196, 214, 217

See also augmented reality
visitor centers *see* interpretive centers
visitor experience 35, 69, 73, 83, 155, 177, 214, 217, 219
visitor management 15, 33–34, 72, 80–81, 162–163, 181, 217
VisitScotland 80–83
volunteer staff, archaeology/heritage 8, 35, 56, 73, 90–92, 100, 156
volunteer tourism 10, 59, 87–102, 211

war 1, 9, 12, 14, 30, 73, 92, 94–95, 110, 128, 130, 140, 175
wear and tear 12, 122–123, 139, 154, 213
websites 100, 128, 159, 163, 217
wine tourism 72
World Bank 55, 153
World Heritage Sites 9–10, 20, 69, 74–75, 80, 106, 115, 117–118, 155–156, 159, 180, 186, 188, 190–191, 206–207, 211
  See also UNESCO
World Tourism Organization (UNWTO) 70, 189, 190

Year of History, Heritage and Archaeology (Scotland) 81–82

For Product Safety Concerns and Information please contact our EU Authorised Representative:

Easy Access System Europe

Mustamäe tee 50

10621 Tallinn

Estonia

gpsr.requests@easproject.com

www.ingramcontent.com/pod-product-compliance
Ingram Content Group UK Ltd.
Pitfield, Milton Keynes, MK11 3LW, UK
UKHW021835140426

5217IPUK00021B/1464